HOW TO ENCOURAGE GIRLS IN MATH & SCIENCE

Joan Skolnick directed the Math/Science Sex Desegregation Project, a federally funded training project for public school teachers in Novato, California. A specialist in curriculum development and in women's studies, she received her Ed.D. from the University of Massachusetts, Amherst, and her M.A. from Columbia University. Her experience in public education includes program administration, research, college and university teaching, and inservice teacher training. She is currently an educational consultant in the Washington, D.C. area.

Carol Langbort is a lecturer at San Francisco State University in both the Mathematics Department and the Department of Elementary Education. Her experience as a mathematics educator includes elementary school and college level teaching, inservice and preservice teacher training, and working with adults who have avoided math. She received her Ph.D. from the University of California, Berkeley, and her M.Ed. from Northeastern University. She was the math consultant to the Novato Math/Science Sex Desegregation Project.

Lucille Day, a math/science education consultant and freelance writer, received her M.A. in zoology and Ph.D. in science and mathematics education at the University of California, Berkeley. Her experience in science and math education includes research, curriculum development, inservice teacher training, and elementary- through-college teaching. She was the science specialist with the Novato Math/Science Sex Desegregation Project.

Strategies for Parents and Educators

HOW TO ENCOURAGE
GIRLS
IN
MATH &
SCIENCE

Joan Skolnick
Carol Langbort
Lucille Day

A SPECTRUM BOOK

Prentice-Hall, Inc., Englewood Cliffs, New Jersey 07632

Library of Congress Cataloging in Publication Data

Skolnick, Joan.
    How to encourage girls in math and science.

    "A Spectrum Book"
    Includes bibliographical references and index.
    1. Science—Study and teaching—United States.
2. Mathematics—Study and teaching—United States.
3. Education of women—United States.   I. Langbort,
Carol.   II. Day, Lucille.   III. Title.
Q183.3.A1S56   507′.1073   82-390
ISBN 0-13-405670-1   AACR2
ISBN 0-13-405662-0 (pbk.)

This Spectrum Book is available to businesses and organizations at a special discount when ordered in large quantities. For information, contact Prentice-Hall, Inc., General Publishing Division, Special Sales, Englewood Cliffs, N.J. 07632.

10 9 8 7 6 5 4 3 2 1

Editorial/production supervision by Louise Marcewicz
and Kimberly Mazur
Interior design by Louise Marcewicz
Page layout by Marie Alexander
Manufacturing buyer: Cathie Lenard

ISBN 0-13-405670-1

ISBN 0-13-405662-0 {PBK.}

Prentice-Hall International, Inc., *London*
Prentice-Hall of Australia Pty. Limited, *Sydney*
Prentice-Hall Canada Inc., *Toronto*
Prentice-Hall of India Private Limited, *New Delhi*
Prentice-Hall of Japan, Inc., *Tokyo*
Prentice-Hall of Southeast Asia Pte. Ltd., *Singapore*
Whitehall Books Limited, *Wellington, New Zealand*

Contents

Preface

Why do many girls avoid math and science? What can parents and educators do to change this? And how can we inspire girls who do like math and science to pursue their interests? *How to Encourage Girls in Math and Science* provides theoretical and practical answers.

Examining the effect of sex role socialization on skills and confidence, the book traces the pattern of girls' involvement with math and science from early childhood through adolescence. It shows how attitudes, parenting and teaching practices, stereotyped play activities and books, peer pressure, and career and family expectations cause girls to question their ability and render them unfamiliar with the math and science skills and concepts they will need.

On the basis of this analysis, the book presents a variety of educational strategies and 69 math and science activities designed to encourage girls. After reading it, you will better understand what needs to be done in our math and science work, at home and in the classroom, to assist girls in overcoming negative influences.

How to Encourage Girls in Math and Science is a parent and educator guidebook. The educational strategies it develops and demonstrates may be applied to any math or science unit or at-home activity. These strategies show you how to promote independence and risk-taking in problem solving, how to utilize manipulative materials and approaches to develop abstract reasoning skills, how to group children during activities to maximize learning and minimize negative peer pressure, and how to use math and science examples to model new sex roles and career possibilities.

The activities, enjoyable and easy to set up, exemplify a multitude of ways to apply these strategies in building critical math and science skills. Even if you know little about math and science, you will be able to help your children develop competence in spatial visualization, working with numbers, logical reasoning, and scientific investigation. Emphasizing problem solving and motivation, the activities are based on sound educational theory and are fun for both boys and girls. Materials such as strategy games, puzzles, kitchen chemicals, imaginary and real animals, and ice cream graphs teach children the pleasures of critical thinking, while enabling them to experience success in math and science.

How to Encourage Girls in Math and Science is the first book that ties particular problems in the socializing of girls to specific resultant learning problems in math and science, while developing compensatory strategies to address each problem. The approach to teaching math and science that emerges from this analysis not only helps girls overcome specific deficiencies, but also presents the subjects from a perspective much closer to that of scientists and mathematicians than is ordinarily found in the school curriculum.

No other book thus far combines such a comprehensive discussion of socialization with a wealth of practical material for parents and educators. Because of the expertise of its authors, the book draws on the disciplines of social science, women's studies, and math and science education. Its rich interdisciplinary approach is rare, yet indispensable to a full consideration of women's role and the full development of their potential in math and science.

The conceptual framework for this book grew out of the work of the Novato Math/Science Sex Desegregation Project under the directorship of Joan Skolnick. The book has its origins in inservice work with teachers and counselors. Each strategy and activity has been tested in classrooms or after-school programs to ensure its success. While all the authors contributed to each section of the book, and teaching strategies were developed collaboratively, Carol Langbort was primarily responsible for writing the math activities, Lucille Day the science activities, and Joan Skolnick the introduction and socialization chapters.

ACKNOWLEDGMENTS

We would like to express our appreciation to the following people who helped make the book possible: the teachers, counselors and students of the Novato Unified School District, who were willing to try new approaches; Sherry Fraser for contributing to the development of ideas as a curriculum specialist to the Novato Math/Science Sex Desegregation Project; Dr. Rita M. Costick for inspiration as a consultant to the Project; Richard Simpson of the Science Graphics Center at San Francisco State University for photography; Dr. Diane Ehrensaft, Dr. Carol MacLennan, and Dr. Randy Reiter for valuable critiques of the socialization chapters. For testing the activities in classrooms, we are indebted to Carol Langbort's students at San Francisco State University, Dorothy Offerman of Sinaloa Junior High School in Novato, Ethel Lagle of West Novato School, Margaret Spurrier Alafi, Director of Twin Pines School in Oakland, and the following Twin Pines teachers: Mary R. Rudge, Sara Blackstock, Sheila Matusek, and Dr. Ted Papenfuss.

HOW TO ENCOURAGE GIRLS IN MATH & SCIENCE

Introduction: Math and Science Divided by Sex

At a local high school, approximately equal numbers of sophomore boys and girls enrolled in science courses. By senior year physics, 150 boys and 46 girls were enrolled. What is the ratio of boys to girls in physics?

Bill and Ann are given a competitive mechanical aptitude test. A high test score will enable a student to enter an apprenticeship program. One test item on time, speed, and distance takes Bill 10 minutes. Ann has trouble diagraming the problem because she can't visualize the relationships. It takes her $2\frac{1}{2}$ times as long as Bill to solve the problem with her approach. How much time has Ann lost?

As a woman, Judy can expect to spend at least 25 years of her adult life working outside the home. She will probably enter one of the traditionally female job categories and can expect to earn about 59¢ for every $1 earned by a man. If both she and Ted (who makes $20,000 a year as an engineer) work 30 years, how much more is Ted's labor worth than Judy's?

A quick computation will tell you that the ratio of boys to girls in that physics class is better than three to one; that Bill will have the competitive edge over Ann for that apprenticeship program since Ann lost fifteen minutes on at least one problem because she had not learned a fundamental spatial skill; and that over the course of her work life, Jane can expect cumulatively to lose almost a quarter of a million dollars compared with Ted.

These math problems are real-life problems. They illustrate a skewed distribution of skills, earning power, and access to education and jobs between the sexes. Such inequities develop through the way each child learns, both in school and outside. Our children and our students are participants in a complex process that equips one sex with math, science, and technical skills indispensible to functioning in the adult world, while it fails to encourage the same development in the other sex. Although the lives of individual women are the most negatively and directly affected, the loss to both sexes is immense.

It is a fact of American life today that family survival is dependent on the abilities and incomes of *all* adults. The kinds of mathematical and technical skills we need to care for our own needs, to be creative, and to survive in the job market escalate daily. At the current pace, computer technology may soon be as basic to literacy as reading and writing. As a society, we cannot afford to inhibit the creativity of over half our population. In these times of economic and environmental crisis, the quality and effectiveness of our social solutions depend on the perspectives that women, as well as men, bring to science and technology.

A decade of changing sex-role awareness has taught us that what plagues one sex usually serves in one form or another to limit the full human potential of the other sex. We stand to enrich the entire world of mathematics and science by asking not why girls can't learn these subjects but why they aren't the sorts of subjects girls want to learn. In approaching our teaching of math and science in ways that encourage the intellectual styles and concerns of girls, we may bridge a gap between technical and social learning. That will benefit us all. In the process, our efforts to understand what makes girls become anxious and avoid math and science will also help many boys with similar difficulties.

Where do discrepancies between the sexes in math and science come from? When do they emerge? Girls and boys start off equal in math and science performance and, as far as we know, in interest. Somewhere in growing up and schooling, the balance shifts. While girls tend to do well in both subjects in elementary school,[1] once courses become optional in secondary school, girls begin a downhill spiral in enrollment, achievement, and interest in math and physical sciences.[2] The decline culminates in inadequate preparation for most college majors and vast underrepresentation of women in all mathematical, scientific, and technical fields of work—in some cases reaching a ninety-nine-to-one ratio of men to women.

Across the nation, slightly more girls than boys are enrolled in our schools. Yet studies reveal that twice as many college-bound senior boys as girls have taken three years of physical science, and similar discrepancies are evident in advanced mathematics enrollment.[3] In a typical school district, boys outnumber girls by more than two to one in most high school physical science courses, three to one in physics.[4] Although girls may outnumber boys in advanced eighth-grade math, by twelfth grade twice as many boys as girls are enrolled in calculus.[5] As a result, relatively few girls are prepared to take the calculus sequence necessary for many college majors.

How important is this for a student interested in liberal arts? It is not surprising that majors in scientific and technical fields such as chemical engineering, computer science, biological sciences, or physics require at least three years of high school mathematics. But it may surprise many to learn that

---

[1]Elizabeth Fennema, "Sex Differences in Mathematics Learning: Why???" *Elementary School Journal*, 75 (1974), 183–90; Thomas L. Hilton and Gosta W. Berglund, "Sex Differences in Mathematics Achievement: A Longitudinal Study," *Journal of Educational Research*, 67 (1974), 231–37; Eleanor E. Maccoby and Carol N. Jacklin, *The Psychology of Sex Differences* (Stanford, California: The Stanford University Press, 1974).

[2]John Ernest, *Mathematics and Sex* (Santa Barbara, California: University of California, 1976); Elizabeth Fennema and Julia Sherman, "Sex-Related Differences in Mathematics Achievement, Spatial Visualization and Affective Factors," *American Educational Research Journal*, 14 (1977), 51–71; National Science Foundation, Directorate for Science Education, *Science Education Databook* (Washington, D.C.: U.S. GPO, 1980); Hilton and Berglund, "Sex Differences in Mathematics Achievement."

[3]National Science Foundation, *Science Education Databook*, p. 31; Education Commission of the States, *National Assessment of Educational Progress: The Second Assessment of Mathematics, 1977–78* (Denver, Colorado: 1979).

[4]Based on data collected by the Math/Science Sex Desegregation Project, Novato (California) Unified School District, 1978–80, Joan Skolnick, Director (Project funded by the U.S. Office of Education); and Lawrence Hall of Science, Equals, *Math Enrollment Patterns in 15 San Francisco Bay Area High Schools* (Berkeley, California: University of California, printout, 1979).

[5]Based on data collected by the Math/Science Desegregation Project, Novato (California) Unified School District, 1978–80, Joan Skolnick, Director (Project funded by the U.S. Office of Education).

you need at least three years of mathematics for majors in business administration, economics, psychology, and architecture;

you need at least two years of high school mathematics to be adequately prepared for programs in art, history, law, and sociology;[6]

you need at least four years of mathematics to work effectively as an airline pilot, veterinarian, or an astronomer;

you need at least two years of mathematics to work effectively as a policeman, a firefighter, a carpenter, or a bank teller.[7]

Students lacking the required math courses enter college with a handicap. They must either make up for their math deficiency (if indeed their college provides high school courses) or immediately limit their choice of major. Even traditionally female courses of study increasingly utilize mathematics and math-related computer technology. Yet course decisions which restrict these adult options are being made as early as adolescence.

Linked to the enrollment problem, learning itself appears to be adversely affected in particular skill areas for girls. The state of California annually tests achievement of sixth and twelfth graders in mathematics. This testing program has revealed a disturbing discrepancy in the ways boys and girls are learning.[8] Specifically,

> girls do consistently better than boys in straight computation and one-step word problems, but boys do consistently better on multiple-step word problems, applications, spatial relationships, more advanced logical reasoning, probability, statistics, and graphs.

Nationwide, seventeen-year-old boys appear to hold the advantage in most of these skill areas.[9] Yet skill areas where girls are weak are critical to advanced math and science problem solving. Echoed by the National Council of Teachers of Mathematics,[10] the California Mathematics Assessment Advisory Committee expressed great concern over the findings:

> Clearly, something is happening in our society, in general,

---

[6]The Mathematical Association of America, *The Math in High School You'll Need for College* (Washington, D.C.: 1978).

[7]The Mathematical Association of America, *You Will Need Math* (Washington, D.C.: 1980).

[8]California State Department of Education, California Assessment Program, *Student Achievement in California Schools, 1977–78 Annual Report, Sex Differences in Mathematics Achievement* (Sacramento, California: 1979).

[9]Education Commission of the States, *National Assessment, 1977–78*.

[10]National Council of Teachers of Mathematics, *An Agenda for Action: Recommendations for School Mathematics of the 1980's* (Reston, Virginia: 1980).

and/or in our instructional programs, in particular, that may be causing nearly half of our school population to perform at a lower level in mathematics than their counterparts of equal capability.[11]

In its recommendations, the committee stated:

> An effort should be made to identify the societal and school factors that may be causing the differences in mathematics performance by boys and girls; and instructional programs, counseling programs, and in-service training programs should be designed to provide maximum and equal opportunities for learning of mathematics by both girls and boys.[12]

The results of early math and science avoidance are clearly reflected in women's position in the labor force. Today, women are entering the job market in unprecedented numbers and staying longer, whether or not they marry. An eighteen-year-old woman who will marry and have two children can expect to spend thirty-four years in the work force; a single woman of twenty can expect to spend forty-one.[13] Most women who work have an economic need to do so. They are single, widowed, divorced, separated, or have spouses whose incomes are insufficient to maintain their family's living standards. Yet women earn on the average only fifty-nine cents for every dollar earned by a man. The dramatically increasing number of American families headed by women are at the very bottom of the income scale, averaging one half the income of male-headed families.

Our occupational distribution is a major reason for the poor earnings of women. The overwhelming majority of women workers are still concentrated in relatively few, lower-paying, lower-status job categories with little chance of advancement. The U.S. Department of Labor reports, "Of 420-odd occupations listed by the 1950 census of occupations, women were employed primarily in 20. That fact was virtually unchanged by 1970."[14] Although women constitute 71% of teachers and 99%

---

[11]California State Department of Education, California Assessment Program, *Student Achievement in California Schools, 1978–79 Annual Report, Mathematics: Grade 12 (Sacramento, California: 1979), p. 2.*

[12]*Ibid.*, p. 3.

[13]U.S. Department of Labor, Bureau of Labor Statistics, *Length of Working Life for Men and Women, 1970*, Special Labor Force Report 187 (Washington, D.C.: U.S. GPO, 1976).

[14]U.S. Department of Labor, Women's Bureau, *The Employment of Women: General Diagnosis of Developments and Issues*, United States Report for OECD High Level Conference on the Employment of Women (Washington, D.C.: U.S. GPO, 1980), p. 6. For data on women's work force participation, see U.S. Department of Labor, Bureau of Labor Statistics, *Employment and Earnings* (Washington, D.C.: U.S. GPO, 1981), and *Perspectives on Working Women: A Databook* (Washington, D.C.: U.S. GPO, 1980).

of secretaries, they make up only 4% of engineers and 1.2% of electricians. Astonishingly, young girls begin limiting their career aspirations to these choices as early as elementary school![15]

The job market, however, progressively favors workers with technical skills based on math and science preparation. The demand in scientific and technical occupations continues to grow rapidly, while the market declines for workers in predominantly female occupations like teaching and humanities. In addition to the greater availability of technical jobs, the average monthly salary for workers with scientific backgrounds is often twice that of workers with humanities backgrounds.

It is apparent that because of changes in the job market, in family patterns, and in the cost of living, there is a critical need for girls to recognize the full range of educational and career possibilities and make informed decisions. But early difficulties with math and science, traditionally sex-typed as male areas, prevent girls from preparing to avail themselves of these work options.

To encourage girls in math and science, we must understand why there is a problem. A great deal of controversy surrounds this question. The very fact that we are dealing with differences between boys and girls has led some researchers to link their search for causes to a *biological* category, sex. A child is born a boy or girl, and somehow that determines whether that child can become a better or worse mathematician. Yet research has failed to demonstrate any innate female inability to learn math and science. Biology cannot explain why such a large majority of women steer away from math- and science-related work.

The most consistent sex differences in mathematical ability have been found in the area of spatial skills. Spatial skills may be defined generally as the ability to manipulate an object or pattern in the imagination. It has been argued that male spatial abilities are responsible for the higher achievement, and even greater interest, of boys in math. However, differences between boys and girls on tests involving spatial skills are relatively small and are not nearly significant enough to explain the *vast* differences in the number of girls and boys who go on in math and science fields.[16] In addition, the differences *among* boys and *among* girls are far greater than the differences *between* the sexes.[17] As we shall see, the development of spatial skills, as all other math/science skills, is related to the kinds of learning

---

[15]Nancy Frazier and Myra Sadker, *Sexism in School and Society* (New York: Harper & Row, Publishers, 1973).

[16]Fennema, *loc. cit.,* p. 183.

[17]*Ibid.*

opportunities we provide each sex. Not surprisingly, training increases spatial ability in girls so that their performance equals or surpasses that of boys.[18]

Many of the studies that claim boys have better mathematical ability do not take into account that girls take fewer math courses.[19] According to Lynn Fox, "when 17-year-old boys are compared to 17-year-old girls the comparison is actually between students with 3–4 years of mathematics, and those with 1–2 years of mathematics."[20] Such comparisons tell us little about children's ability. When children take an equal number of courses, sex differences in math performance often tend to diminish or disappear.[21]

But we must look beyond formal course work to paint the whole picture of girls' difficulty in math and science. A recent very controversial study by Camilla Benbow and Julian Stanley[22] erroneously assumed that boys and girls who had taken equivalent course work had had the same formal educational experiences and training in math. When they discovered, then, that the boys in their sample had higher "aptitude" scores, they supported the theory that boys have superior mathematical reasoning ability. This questionable conclusion was widely publicized in the press,[23] although we know that children learn math and science concepts *outside of school* and that *even in the same classroom*, boys and girls often learn different lessons. All this learning affects their math achievement.

We cannot say that biology determines girls' math and science destiny any more than it determines why boys do not do housework or become secretaries and nurses. On the contrary, the evidence suggests that the problem has far less to do with whether you are born a boy or girl than with what society makes of that fact. For this reason, we will learn a great deal more about helping our children reach their full potential by shifting

---

[18]E. H. Brinkmann, "Programmed Instruction as a Technique for Improving Spatial Visualization," *Journal of Applied Psychology*, 50 (1966), 179–84.

[19]Elizabeth Fennema, "Influences of Selected Cognitive, Affective, and Educational Variables on Sex-Related Differences in Mathematics Learning and Studying," in National Institute of Education Papers in Education and Work No. 8, *Women in Mathematics: Research Perspectives for Change* (Washington, D.C.: U.S. GPO, 1977), pp. 79–136.

[20]Foreword, NIE Papers, *Women in Mathematics*, p. iii.

[21]Elizabeth Fennema and Julia Sherman, "Sex-Related Differences in Mathematics Achievement and Related Factors: A Further Study," *Journal for Research in Mathematics Education*, 9 (1978), 188–203.

[22]C. P. Benbow and J. C. Stanley, "Sex Differences in Mathematical Ability: Fact or Artifact?" *Science*, 210 (1980), 1262–64.

[23]"At Mathematical Thinking, Boys Outperform Girls," *The Washington (D.C.) Post*, December 5, 1980); "The Gender Factor in Math," *Time*, 116 (1980), p. 57; "Do Males Have a Math Gene?" *Newsweek*, 46 (1980), p. 73.

our focus from "sex" to "gender." While sex is a biological given, gender is a social, cultural, and psychological creation. Children learn that being a boy or girl makes a difference in what they are expected to be, do, think, and feel. This learning process is called gender socialization. Differences in the treatment and expectations of girls and boys from infancy on influence the kinds of skills each sex develops through play, how confident children feel as learners, and even what intellectual risks they are willing to take. As a result of gender socialization, girls learn to approach math and science with greater uncertainty and ambivalence than boys, with inadequate practice and unfamiliarity in particular skill areas (like spatial skills) and, more generally, with conflicts about competence and independence. Gender mediates what and how we learn.

Since the reasons girls do not continue in math and science have to do with both social pressures and math/science skills, this book will focus on both. First we examine gender socialization at home and in schools as it influences girls' learning. Based on that analysis, we identify skill-building and confidence-building strategies to encourage girls. Finally we develop kindergarten through eighth grade math and science activities to use with girls and boys at home or at school.

Two points should be noted about our discussion. Different ethnic and economic groups vary in their socialization practices. These variations may well affect girls' attitudes and skills in math and science, although little research is available to tell us how. Our purpose here is to provide an overview of the critical factors affecting girls' math and science learning. Further research should improve our understanding of how these factors operate and vary in particular portions of our population. Secondly, we focus on improving individual attitudes and skills. Other factors also operate to limit opportunities. The range of jobs available, and the incentives and training programs offered to women are affected by forces shaping the American economy. Social support for working women, such as child care programs, are affected by political priorities. Issues such as these need to be addressed. Our concern here, however, is with how children learn and how adults can encourage learning to increase girls' access to the range of jobs that now exist.

As parents, teachers, administrators, and friends, we may be unwitting accomplices in developing math and science inequities. But it is precisely because human agency created the problem that we can solve it. This book is a contribution to that effort.

I ECHOES FROM CHILDHOOD

Parent, Teacher, and Child: Learning Confidence

1

Working through complex mathematical problems and formulating scientific hypotheses involve a great deal more than skills application. They are as much a matter of taking risks, making mistakes without becoming devastated, persisting in independent work, and trusting our ability to solve problems and effect changes in our environment.[1] Women, more often than men, feel inadequate at problem solving and underestimate their intellectual abilities.[2]

Helplessness has historically been a hallmark of femininity among the more affluent classes. Although today they are mixed with ambiguous metaphors of liberation, images of feminine helplessness still pervade our culture and affect our feelings and behavior. Textbooks and media still portray baffled women relying on men to compute bills, make simple repairs, and advise them on all practical uses of science in their lives—from nutrition to cleaning fluids. In social interaction, helplessness in women is still often expected and rewarded. This

---

[1]For further discussion of risk taking in mathematics, see Sheila Tobias, *Overcoming Math Anxiety* (New York: W. W. Norton & Co., Inc., 1978).

[2]Kay Deaux, *The Behavior of Women and Men* (Monterey, California: Brooks/Cole Publishing Company, Inc., 1976), pp. 38–39.

stereotype contributes to both women's lack of confidence and their lack of practice in problem solving.

As early as infancy, we begin to develop trust in ourselves as learners. As children grow, the expectations and practices of parents, teachers, counselors, and peers can nurture the growth of confident, independent learners or promote fears and conflicts. Indications are that girls and boys are taught very different lessons. In the following pages, we will explore adults' sex-linked expectations of children and how they affect children's attitudes towards math and science learning.

FIGURE 1-1. Novato Children's Center, Novato, California.
Photo by Ron Moncayo.

PARENT AND CHILD

Which would you rather host—a birthday sleepover party for twelve of your nine-year-old son's favorite boyfriends, or one for his sister's favorite girlfriends? In the first case, you might be tempted to hire a proxy and leave town, taking up rug and breakables on the way out. For the girls' party, you might anticipate through-the-night giggling but probably would not consider ear plugs a requirement.

All of us have expectations of children that may not reflect the personalities of the specific girls and boys we meet. Understandably, our expectations affect our behavior. The parent who removes valuables and buys ear protection prior to a son's birthday party delivers a message to the boys: they are expected to be rowdy, disturbing, and naughty. The girls, on the other hand, may be silly, but they are harmless. If such messages about personality are repeated often enough, they become a part of children's perceptions about themselves. Over time, children may begin to act the way they are "supposed to be." Their behavior, in turn, reinforces adult convictions that boys and girls are simply different by nature.

But adults begin so early to respond to their children in a sex-typed fashion that a child's nature is always molded by gender expectations.[3] In our culture, the sex of a child is considered so important that it preoccupies our imagination long before the birth of the baby. "Boy or girl?" is the first question asked after delivery, the first label attached to the child on the birth announcement, and the first piece of information sought by friends and relatives in buying gifts. The answer to the boy-or-girl question not only decides the proper color of the baby blanket but also affects people's perceptions about the baby's very nature—its size, activity, attractiveness, and even its future potential. Within hours after birth, newborn babies of the same weight, length, and general characteristics are described differently by their parents. Girls are seen as smaller, softer, more fragile, weak, and beautiful; boys as stronger, larger, more alert, coordinated, aggressive, and even athletic.[4]

From birth baby girls and boys do show some differences that can influence how their parents react to them. But what parents make of sex differences is crucial, especially since many sex stereotypes actually contradict biological realities. Unlike the stereotyped description of newborns, for example, boy infants tend to be weaker and more fretful. Perhaps as a result, parents handle and rock them more than girls during the first few months. However, by six months, some studies show parents are already reversing this pattern and weaning boys for a more independent life.[5] Often parents' own sex role expectations and behavior predispose them to encourage those traits in their babies that fit sex stereotypes and to discourage those that don't.

Early differences in the treatment of girls and boys can initiate enduring learning patterns. To develop confidence,

---

[3]Michael Lewis, "Parents and Children: Sex Role Development," *School Review*, 80 (1972), 229–40.

[4]J. Z. Rubin, F. J. Provenzano, and Zella Luria, "The Eye of the Beholder: Parents' Views on Sex of Newborns," *American Journal of Orthopsychiatry*, 44 (1974), 512–19.

[5]Lewis, *loc. cit.,* p. 234.

independence, and intellectual skills, babies need to learn that they can affect their world. In crawling, climbing, talking, and touching, they explore, experiment, and build upon their successes. During these very important first years, gender expectations help boys build trust in their ability to master their environment. In comparison, girls may learn to depend more on their proximity to adults and less on their own abilities. Some studies[6] have found that parents:

> are more apprehensive about girls' well-being and protect them more;
>
> tend to direct boy babies away from mother starting at about six months, thus creating more opportunities for independent exploration, physical activity, and problem-solving;
>
> keep girls closer to home during play and assist them more in their activities; and
>
> encourage and maintain the close physical contact of early infancy for a longer time with daughters.

The less girls are encouraged to develop the skills and confidence to cope with their environment, the more fearful they are likely to become when "abandoned" to their own devices.[7] In one study, girls were already showing signs of such dependency by thirteen months.[8] Baby girls returned to their mothers more quickly and frequently when put down, and when a barrier was placed between mother and child, the girls became more upset and less aggressive in attempting to cross it.

These studies suggest a general pattern of dependency training for girls. Although this area of research is still controversial,[9] such a pattern is consistent with strong evidence

---

[6]See particularly, Michael Lewis, "Parents and Children: Sex Role Development," *School Review*, 80 (1972), 229–40; Michael Lewis and Marsha Weinraub, "Origins of Early Sex-Role Development," *Sex Roles*, 5 (1979), 140–42; Lois Wladis Hoffman, "Early Childhood Experiences and Women's Achievement Motives," *Journal of Social Issues*, 28 (1972), 129–55.

[7]Hoffman, "Early Childhood Experiences," pp. 129–55.

[8]Susan Goldberg and Michael Lewis, "Play Behavior in the Year-Old Infant: Early Sex Differences," *Child Development*, 40 (1969), 21–31.

[9]There is still much controversy about just how boy and girl babies are treated differently and to what extent such differences affect children's behavior as they grow. We know that adult expectations of newborns are stereotyped and that by preschool there are differences in adult treatment of boys and girls. Although studies on the intervening years are indecisive, it is reasonable to believe sex differences in the treatment of the young are consistent with how adults see their newborns and treat their school-age children. For a review of issues and research in this area, see Eleanor Maccoby and Carol Jacklin, *The Psychology of Sex Differences* (Stanford, California: Stanford University Press, 1974).

about adult stereotyping of newborns and treatment of pre-schoolers.[10] If reinforced later by schooling, peer interaction, words and pictures, girls' growing distrust of their own abilities may develop into serious intellectual self-doubt and learning anxiety. Struggle with these issues can impede girls' development of their talents, particularly in unfamiliar skill areas, like math and science.

Another early parenting practice can also have serious implications for the learning styles of boys and girls. In our society women are the primary caretakers of infants and young children. Because of this, the earliest identification for both girl and boy babies is with mother, a woman. In fact, infants first experience themselves as merged with their world and with mother, the closest object in their world. As they mature physiologically, babies of both sexes begin a long process of separating themselves from mother. For girls this separation process is never as complete as it is for boys, because mother is of the same sex.[11]

One of the earliest and most emotionally charged learning tasks in a child's life is gender identification. When a girl learns what it means to be female, she learns to be "like mother." Her adult role model is usually ever-present in her environment, to be imitated and to provide immediate feedback. For boys, however, gender identification is a more abstract task. Without fathers as close by daily, boys can neither rely as much on direct imitation of specific behaviors nor on close personal interaction and feedback from a male role model. A boy learns that he is supposed to be "not like mother." Largely from such negative cues and from cultural stereotypes, a boy must figure out the principles of masculinity.[12]

Thus, gender identification presents different learning problems for boys and girls to solve. Psychologist David Lynn argues that in the process boys accumulate practice at important kinds of problem solving and develop a learning style that primarily involves (1) defining the goal, (2) restructuring the situation, and (3) abstracting principles. The girls' task, on the other hand, involves "learning the lesson as presented" and

---

[10]L. Cherry, "The Preschool Teacher-Child Dyad: Sex Differences in Verbal Interaction," *Child Development*, 46 (1977), 532–36; L. A. Serbin, K. D. O'Leary, R. N. Kent, and I. J. Tonick, "A Comparison of Teacher Response to the Pre-Academic and Problem Behavior of Boys and Girls," *Child Development*, 44 (1973), 796–804.

[11]For a full discussion of how women's role in parenting results in female/male emotional differences, see Nancy Chodorow, *The Reproduction of Mothering, Psychoanalysis and the Sociology of Gender* (Berkeley, California: University of California Press, 1978).

[12]David B. Lynn, "Determinants of Intellectual Growth in Women," *School Review*, 80 (1972), 247.

promotes a style based more on (1) personal relationship and (2) imitation.[13]

We should not conclude that girls do not learn to abstract. Even the researchers who test abstract ability cannot precisely agree on its meaning. We can say that as they grow, girls' abstract reasoning develops in a different context (highly interpersonal) and focuses on a different content (people and relationships). As a consequence, girls are less practiced at particular kinds of analytic tasks utilized in mathematics and science.[14] Their familiar social style of learning and their motivation to learn people-oriented subjects are incompatible with math and science as they are currently taught.

Other intellectual and emotional lessons of early childhood are also transferred to later learning. Girls may develop early proficiency at language skills and at rote learning, which facilitates memory. But they do not develop as fixed a sense of their own boundaries,[15] nor are they as motivated to assert their independence. Used to learning in close affiliation with others, they may later feel tentative about their own abilities and overly attached to teachers and formulae for direction in their work.

Boys' autonomy gives them greater practice at some analytic tasks but deemphasizes language and exacts a high emotional toll. Since they learn gender in closer proximity to mothers than fathers, they learn early to emphasize their separateness. Later, they may have greater difficulty developing a strong sense of connectedness with others.[16]

As men take a more active daily role in parenting their children, the poles of male and female learning styles might not be so far apart, and the resultant skills might be more equitably distributed. In the future, each child may be more able to draw the best lessons of both early girlhood and early boyhood, combining independence, emotional breadth, language, and analytic abilities.

In the meantime, we as educators and parents can take some steps to address the learning problems children develop currently. To increase girls' achievement in math and science, we must encourage the growth of intellectual self-confidence, develop problem-solving skills, and build on girls' verbal and interpersonal strengths in the learning process. Strategies to accomplish this will be described in chapter four.

---

[13]*Ibid.*, p. 248.
[14]*Ibid.*, pp. 250–52.
[15]Chodorow, *op. cit.*
[16]*Ibid.*

Children spend more time with their teachers than with any other adults except parents. They see themselves reflected in their teacher's eyes; what their teacher thinks counts heavily in their world. Outside the schoolground, young children invoke their teacher's opinions as the last word on almost every subject.

It is not surprising that teachers' expectations can affect children's achievement. In one famous elementary school study, teachers were told at the start of the school year that certain of their students had high potential. Although the names of these students had in fact been randomly selected, the teachers' high expectations so influenced instruction that these children actually did better and were reported as more curious and interested than their classmates.[17]

Teachers' expectations are communicated to children in myriad ways, not only through what they say explicitly but also through what they do not say, what they do, and whom they call on. Indirect or covert messages constitute a hidden curriculum which is sometimes more powerful than the lessons in the textbooks, and many times reinforced by those lessons.

In theory, most teachers believe education should be a liberating and democratic influence: education should give each child the same opportunity to reach full potential regardless of race, sex, or economic background, so that he or she may earn a place in life based on merit and hard work. But even the most egalitarian-minded teachers sometimes unintentionally respond to children in a sex-typed fashion counter to those principles. The record shows that instead of altering traditional sex stereotypes, schooling tends to reinforce the sex-role lessons of infancy and early childhood.[18] However, supported by the federal law against sex discrimination in education (Title IX), many teachers are currently trying to promote fairer educational practices by becoming more aware of sex biases in the hidden curriculum.

How do teachers treat girls and boys differently? Do they praise or reprimand one sex more than the other? Do they encourage one sex to try harder? And if so, how do these

---

[17]Robert Rosenthal and Lenore Jacobson, *Pygmalion in the Classroom: Teacher Expectation and Pupils' Intellectual Development* (New York: Holt, Rinehart & Winston, 1968).

[18]See Nancy Frazier and Myra Sadker, *Sexism in School and Society* (New York: Harper & Row, Publishers, 1973).

differences affect our children? To find out, educational researchers have observed and tabulated the kinds of classroom interactions teachers have with children of each sex during a normal day. They discovered that boys receive more classroom attention than girls.[19] They are called on a greater number of times, are asked more direct, open-ended, complex, and abstract questions, and receive more detailed instruction. Since individual attention from the teacher is a form of recognition and helps probe a child's thinking about the subject matter, children who receive more of it are at an advantage. Over the course of their schooling, boys benefit from an interactive teaching style which promotes achievement.

A national study to determine why girls discontinue mathematics documented differential treatment of boys and girls even in high school geometry classes.[20] While boys were called on, spoken to, and asked more questions, the only preferential communication with female students was social and not intellectual—teachers smiled or laughed more with the girls than they did with the boys. Thus, even when boys and girls take the same number of math classes, their training experiences differ. Studies that conclude that girls are genetically inferior in math ability because their test scores fall below those of boys with the same course work ignore the cumulative effect of this differential treatment.

Children's self-esteem is affected not only by the amount but also by the kind of teacher attention they get. In turn, the rewards and punishments dispensed by teachers reflect sex-role expectations. As part of the male sex role, adults expect higher achievement, as well as more aggressive, independent, and unruly behavior. Boys are scolded eight to ten times more often than girls.[21] This negative attention is mostly directed at disruptive social behavior and disobeying the rules of form—for example, turning in messy papers. On the other hand, almost all the praise given boys concerns their intellectual qualities and academic work.[22] Through their teachers' reactions to them, boys learn to see themselves as, perhaps, naughty, undisciplined, rebellious, or unmotivated, but as essentially intelligent and capable. The hidden curricular message for boys reads,

---

[19]For a review of teacher-student interaction research and a workbook for teacher education, see U.S. Department of Education, Women's Educational Equity Act Program, Non-Sexist Teacher Education Project, Myra Sadker and David Sadker, *Between Teacher and Student: Overcoming Sex Bias in Classroom Interaction* (Newton, Massachusetts: Educational Development Center, 1979).

[20]Jane Stallings and Anne Robertson, *Factors Influencing Women's Decisions to Enroll in Advanced Mathematics Courses,* final report for the National Institute of Education (Menlo Park, California: SRI International, 1979), p. 63.

[21]Sadker and Sadker, *op. cit.*, p. 18.

[22]*Ibid.*, p. 28.

"If you would only behave yourself and *try*, you would succeed."
When boys do not succeed, teachers tend to attribute their
failure to lack of effort rather than lack of skill,[23] and they are
instructed to try again.

Thus, despite the reprimands, boys' disruptive behavior
has its rewards: it captures adult attention and implicitly pro-
tects the intellectual ego. After all, no test is a fair test of ability
if the child didn't try. Although elementary school boys may
have problems with emotional adjustment to school life, teacher
feedback helps them develop a more resilient intellectual self-
confidence than girls: one poor test performance does not as
readily cause them to question their capabilities, and failures
are more readily blamed on external circumstances, such as a
bad teacher or an unfair allegation of cheating.[24] In fact, boys
tend to estimate their chances of success higher than girls, even
in skill areas where girls actually do better.[25] As a result, they are
often more willing to take the risks that help them develop
know-how.

FIGURE 1-2. Making Mountains, Chapter 8. Twin Pines School, Oakland, California.
Photo by Richard Simpson.

Most elementary teachers will tell you they have more
occasion to praise little girls than little boys. But the qualities

[23]Carol Dweck, William Davidson, Sharon Nelson, and Bradley Ehna,
"Sex Differences in Learned Helplessness," *Developmental Psychology*, 14 (1978).
[24]Deaux, *op. cit.,* p. 32.
[25]*Ibid.,* pp. 38–39.

rewarded in the two sexes are very different. While boys are usually praised for intellectual work, girls are mostly praised for behaving properly and obeying rules of form. They are encouraged to be compliant but not necessarily to be creative, autonomous, or analytic. As a result, they learn they are pleasing but not necessarily that they are capable. In fact, sex-role expectations of girls are consistent with elementary teachers' descriptions of the model student: cooperative, calm, obedient, considerate, neat. Teachers' praise, then, puts pressure on girls not to deviate from their role.[26] Furthermore, at school, as at home, the teacher is usually a woman. As girls remain close by their teachers physically and emotionally, the dependencies of infancy and early childhood are reinforced.

Teachers also criticize girls for different things than boys and give them different feedback when they fail. While boys are scolded for being disruptive, ninety percent of the criticism given girls directly concerns their intellectual inadequacy.[27] Because poor performance by a girl is not usually attributed to lack of effort, teachers, like parents, are more apt to respond by performing the difficult task for her, consoling her for not doing well, or congratulating her for having tried in the first place. They instruct girls less often than boys to try again on their own. By inference, girls learn to see their failures as the result of innate inability. In essence, they learn to view themselves as helpless. For girls, the hidden curricular message reads, "You've followed all the rules and haven't succeeded. Perhaps you're just not good at it. But try not to worry, dear."

To test the effect of such feedback on children's self-evaluation, fifth grade children were set to work on word puzzles and given different feedback. Most girls and boys receiving the kind of feedback boys typically get attributed their puzzle failures to lack of effort. Most girls and boys receiving the more female feedback interpreted their failures as lack of ability.[28]

As we shall see, these messages about ability are often reinforced in informal learning outside of school, and, likely, within the family itself. Often high-achieving girls with less accomplished brothers end up feeling that their brothers are *really* smarter. In family mythology, the girl succeeds because she works so hard and follows the rules. But it is assumed that her brother, who gets through with little effort, would excell "if only he'd put his mind to it."

Thus, in general, males "are credited with ability when they succeed, while failures are attributed to external circum-

[26]Patrick C. Lee and Nancy B. Gropper, "Sex-Role Culture and Educational Practice," *Harvard Educational Review,* 44 (1974).

[27]Sadker and Sadker, *op. cit.,* p. 28.

[28]Dweck, "Sex Differences in Learned Helplessness," p. 274.

FIGURE 1-3. Making Mountains, Chapter 8. Twin Pines School, Oakland, California. *Photo by Richard Simpson.*

stances" or motivational factors. For females "these patterns are nearly reversed."[29] The long-range cost to girls' self-confidence is incalculable. Their early lead in school drops off, and by the end of the elementary years many girls have begun to limit their aspirations and to fear intellectual risks.[30] In math and science particularly, they begin to attribute their successes to luck and not to skill.[31] As children continue through school, boys do increasingly better, while girls' self-esteem, their opinion of their sex, and their scores on standardized achievement tests decline.[32] At the same time, assertiveness, indepen-

---

[29]Deaux, *op. cit.,* p. 41.

[30]Vaughn Crandall and Alice Rabson, "Children's Repetition Choices in an Intellectual Achievement Situation Following Success and Failure," *Journal of Genetic Psychology,* 97 (1960), 161–68. Crandall and Rabson found that elementary school girls preferred to repeat puzzles they had already completed rather than tackle the new ones preferred by their male classmates.

[31]Sharon Churnin Nash, "Sex Role as a Mediator of Intellectual Functioning," in *Sex-Related Differences in Cognitive Functioning,* ed. Michele Andrisin Wittig and Anne C. Peterson (New York: Academic Press, 1979), pp. 263–302.

[32]For a summary of the effects of sex stereotyping in schooling, see Myra Sadker and David Sadker, *The Report Card: The Cost of Sex Bias in Schools* (Washington, D.C.: The Mid-Atlantic Center for Sex Equity, The American University, 1980).

dence, and self-confidence, undermined in female socialization, become increasingly important to academic success.

IMPLICATIONS FOR LATER LEARNING

In this chapter, we have seen how parenting and teaching practices encourage girls' distrust in their own abilities. These practices promote reliance on adult feedback, a preference for interpersonal and verbal skills, and limited practice at independent abstract problem solving of certain kinds. How will these issues affect math and science learning? Distrusting one's abilities and relying on others can have a negative effect on *any* learning task. But they are likely to be most problematic in subjects which are out-of-role for girls and thus have few female role models, arouse negative peer pressure, and build on less familiar skills such as spatial, mechanical, and abstract reasoning.

Mathematical problem solving and scientific investigation are lonely, risk-taking procedures. Although in the real world, scientists often need to work together to solve a problem, in the classroom, math and science are taught in an asocial manner. Math and science problems are assigned to students to complete individually; they are rarely considered proper themes for essays, group discussions, or debate. In geometry, chemistry, and physics, numbers and symbols take on a life of their own, increasingly removed from what they represent in ordinary human language and activity. Girls, with well-cultivated interests in people and strong verbal skills, are both in unfamiliar terrain and very much on their own.

Within just one advanced algebra problem, the solitary learner must manipulate a series of complex symbols, choose and apply different operations, and work through many steps in logical reasoning, each one building on those preceding. There is little communication, direct guidance, clear-cut feedback, or external reward until and unless the student reaches the end of the problem. Teacher approval, important to girls, is dependent on finding the one right answer. But to get there, the learner must trust her capacity to reason and be able to provide her own subjective evaluation of how she's doing. Since girls learn early to question their own abilities and depend on interaction with others, it is not surprising that more girls than boys find mathematics an anxious experience.

Similarly, formulating hypotheses is risk-taking, independent work requiring the learner to generalize and abstract, abilities encouraged early in boys. Since hypotheses are really educated guesses, conducting a successful experiment depends on the capacity to weather mistakes. It is therefore confidence, and not just skill, that permits the experimentation necessary for success.

Yet female socialization creates considerable anxiety about the possibility of being wrong. Many girls reportedly feel they've failed, for example, if they bring home a B on a test rather than an A. The imperfect grade can represent more than the misunderstanding of a particular concept. It can confirm a girl's belief that she is not *really* smart; it can represent the withdrawal of a teacher's approval; it can be interpreted as a personal failure, since by this time pleasing has become so important to girls.

In addition, women are more likely to predict failure at the outset. If you predict failure on a math or science problem, any mistake threatens to confirm the prediction, and you might give up along the way. If you believe your successes are really a result of luck, as more women do, you might tend to view your own steps in logic as random. There are many possible results: avoidance, blocking, reluctance to make and test guesses, hesitance to plunge into a new kind of problem, fear of proceeding without teacher direction. High school teachers and math anxiety clinics[33] have observed many of these learning problems to be more characteristic of women.

There is a mystique about mathematics that implies that either you have what it takes or you don't. This notion is all too compatible with social messages about female inability. Many girls believe they simply lack the particular talent, rather than the practice, that permits success. To dissolve the mystique, girls need to do what boys do—practice. As they experience success, they will gain credibility with themselves as math and science learners. Parents and teachers can help them by redirecting the focus of math and science activities so that confidence in problem solving is primary.

In chapter four we describe strategies to encourage independence and risk-taking in solving problems. Our major objective is to empower the student by helping her believe in her own ingenuity, creativity, and analytic processes. She may then utilize the critical thinking tools she develops to solve problems in many settings.

All of us, adults and children, males and females, can use more confidence. Identifying some of the issues that produce self-doubt in girls may help us deal with similar problems in ourselves and boys. As sensitivity informs our practices, we not only build skills but also increase our pleasure in learning.

---

[33]See, for example, Tobias, *op. cit.*

2 Toys, Games, and Books: Building Skills

Have you ever overheard a conversation that goes like this:

> She asks him to take over some domestic activity—to clean the house, host the morning's child co-op, or prepare dinner for guests.
>
> Flustered, he responds, "I can't—I don't know what to do!"
>
> She replies, "What do you mean you don't know what to do? You've got a mind. Figure it out!"
>
> "Okay, okay, I'll do it! But I'm sure I won't be able to do it right. Why don't you just show me how?"
>
> "You're hopeless. I might as well just do it myself!"

Often, quite competent men who manage businesses, analyze foreign policy, or work with complicated machinery plead that I-don't-know-where-to-begin feeling when faced with a relatively simple exercise in child care or domestic science. Few would attribute this helplessness to genetic inability, although skeptics dismiss such male protestations as an excuse to avoid unpleasant work. Nevertheless, the cry of bewilderment expresses a truth about confidence: people feel insecure, unable, and dependent when attempting to do things at which they have had no practice. All the intellectual tools they would normally use to solve problems desert them, and simple tasks seem overwhelming. When the task is gender-related, as in this case, an irritable inner voice says, "I shouldn't be doing this sort of thing at all!" Consequently, they resist the practice that is so intricately linked to confidence.

Throughout their development, girls get little exposure to math and science concepts, and even less practice at actually doing math and science. We have previously seen how gender conditions girls' learning, adversely affecting self-confidence and accustoming girls to learning styles and skills not favored in math and science. Opportunities to practice less encouraged styles and skills could break the anxiety-avoidance cycle at almost any point. But through sex-stereotyped toys, games, books, and subject matter, girls are channeled away from the experiences that could give them confidence as math and science learners.

TOYS AND GAMES

Young children learn important cognitive abilities through play. Yet boys and girls play with different kinds of toys and at different games and activities. Almost as soon as a child is born, adults contribute to the development of toy preferences, setting the parameters of toy choices long before anyone knows what

FIGURE 2-1. *Photo by Richard Simpson.*

the individual child prefers. Parents not only buy different toys for their sons and daughters, they also play with their children differently.[1] By the age of three, children have already begun to pressure one another to play with sex-appropriate toys.[2] Later, peer sanctions against tomboys and sissies create painful struggles for children who commit the crime of preferring the wrong play activities.

Sex differences in play can have repercussions for children of both sexes. For girls, it can set the stage for later difficulty in math and science by limiting familiarity with basic concepts and important skills such as spatial visualization, the ability to manipulate an object in the imagination. For boys, it can result in weaker social skills and unfamiliarity with nurturing roles.

Girls tend to play with toys that favor verbal, interpersonal, and fine-motor skills. In playing house, girls simulate families, practice nurturing and caretaking, make up stories, and talk to the dolls and to each other. They use objects to explore emotions and relationships. They rarely take dolls apart to see how they work or throw dolls in the air and position themselves according to the angle at which they think the doll will land— play activities which involve math and science concepts.[3] Fine-motor activities, such as drawing and sewing, tend to be sedentary and structured. They keep girls close to home and do not promote physical maneuvering through space or unanticipated independent problem solving.

Girls' games generally involve smaller, more intimate groups than boys' games, and, often, concern with social relationships is as fundamental as the content of the game itself. These concerns are reinforced by parents. Studies show that parents playing the same game with sons and daughters stress achievement with sons (winning the game) and "just being together" with daughters.[4]

Instead of emphasizing relationships with people, the toys and games of little boys more consistently involve them with objects: finding the relationship of one object to another, taking things apart and fitting them together, manipulating objects in space, and visualizing and grouping objects. These skills are a major part of such activities as building with blocks,

---

[1]Television interview with Jeanne Block, University of California, Berkeley. "The Pinks and the Blues," *Nova,* WGBH-TV, Boston, viewed September 30, 1980.

[2]Lisa Serbin, Jane M. Conner, Carol J. Burchardt, and Cheryl C. Citron, "Effects of Peer Presence on Sex-Typing of Children's Play Behavior," *Journal of Experimental Child Psychology*, 27 (1979), pp. 303–9.

[3]For further discussion of mathematics and play, see Sheila Tobias, *Overcoming Math Anxiety* (New York: W. W. Norton & Co., Inc., 1978), pp. 91–95.

[4]Television interview with Jeanne Block, "The Pinks and the Blues."

constructing model airplanes, repairing bicycles, setting up and operating electric trains, and experimenting with microscopes and erector sets. Boys' toys, particularly climbing and riding toys, also involve more physically active navigation through space.

Outside of school, boys' interaction with their environment and with one another in sports and games is bound up with what Sheila Tobias calls "street mathematics."[5] Batting averages, interception of a ball in the air, trajectories of balls, and scoring all involve calculations of numbers, time, speed, or distance. Boys playing with balls, bikes, and billiards are actually doing experiments and learning the principles of physics. Boys use math and science concepts to negotiate with one another over game rules and plays. These concepts become basic to their language of social interaction. Later, many of the specific examples they encounter in studying subjects like physics actually come from the world of boys' play.

Learning where to hit the cue ball to produce the proper angle or how to speed a bike through a jungle gym to avoid a crash not only instructs boys in physics, it also develops spatial imagination. Remove a boy from the billiard set and the jungle gym, and often he produces his own imaginative theater with accompanying sound track: he leans forward, extending his arms into the movement of the cue ball as it angles off the cue stick; he grips invisible handlebars to maneuver the bike through the twists and turns of its obstacle course.

Spatial imagination, or visualization, includes the many skills involved in maintaining spatial orientation and manipulating objects in the mind—for example, picturing how an object would look turned upside down, sliced down the middle, or partially rotated. Such skills are essential in sciences and math. They form the very framework of geometry, a subject in which many bright, high-achieving girls begin to develop the difficulty that eventually deters them from mathematics. Spatial ability is considered so important to another area, mechanical aptitude, that it is often given special emphasis in aptitude testing for apprenticeship, vocational, and trade programs.

Starting in adolescence, females tend not to do as well as males on tests involving spatial abilities.[6] Some have argued that this demonstrates a biological barrier to women's achievement in mathematics.[7] Although some biological factors may

---

[5] Tobias, *loc. cit.,* p. 167.

[6] Elizabeth Fennema, "Spatial Ability, Mathematics and the Sexes," in *Mathematics Learning: What Research Says About Sex Differences,* ed. E. Fennema (Columbus, Ohio: ERIC Center for Science, Mathematics, and Environmental Education, College of Education, Ohio State University, 1975).

[7] For a review of literature on spatial abilities and sex-related differences in mathematics, see also Julia Sherman, "The Effects of Biological Factors on

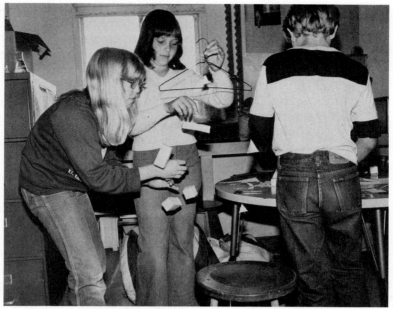

FIGURE 2-2. Make A Mobile, Chapter 5. Twin Pines School,
Oakland, California.
Photo by Richard Simpson.

be operating, research indicates that sex-role socialization is the
major deterrent to the development of girls' spatial abilities.
Beginning with childhood toys and games, girls have limited
practice in spatial visualization.[8] But just as boys' verbal skills
improve with practice, research shows that girls' spatial skills
improve with practice. When teachers involve preschool girls
and boys in concentrated, daily manipulation of blocks, a male-
preferred activity, the spatial skills of all the children improve,
with girls benefiting the most.[9] Even in adulthood, intervention
programs can equalize women's spatial task performance with
that of their male peers.

 Since verbal skills are encouraged early in girlhood, it may
be that girls develop a preference for verbal analysis at the
expense of less familiar spatial approaches to problem solving.[10]

Sex-Related Differences in Mathematics Achievement" in *Women and Mathematics: Research Perspectives for Change,* NIE Papers on Education and Work No.
8 (Washington, D.C.: U.S. GPO, 1977), pp. 137–91.

 [8]Jane M. Conner and Lisa Serbin, "Behaviorally Based Masculine- and
Feminine-Activity-Preference Scales for Preschoolers: Correlates with Other
Classroom Behaviors and Cognitive Tests," *Child Development,* 48 (1977),
1411–16.

 [9]Lisa Serbin, presentation to the Institute for Equal Early Education,
June 12, 1980, Loveland, Ohio; sponsored by the Non-Sexist Child Development Project, the Women's Action Alliance, Inc.

 [10]Sherman, *loc. cit.,* p. 167.

They then tend to use verbal strategies, even when spatial strategies, such as diagraming, might work better. Observations of math teachers corroborate this theory. They note that girls try to "talk their way through" math problems rather than visualizing numerical relationships and drawing pictures or graphs.

If girls have their own compensatory ways of approaching problems, why are spatial visualization skills so essential? First, spatial approaches often make problem solving a lot easier and save time and frustration. Second, verbal approaches are not always effective. Additionally, in our day-to-day experiences, the amount of data with which we must deal is so extensive that it is continually represented to us pictorially. Check the newspapers and media. Rarely a day goes by when the news anchor is not framed by the backdrop of a graph or other pictorial representation of the latest Gallup poll and the day's economic story. Such visual messages give us access to critical social information. To those of us not practiced in spatial literacy, graphs can mystify current events and deprive us of some of our power to make informed decisions.

THE SEX OF SUBJECTS

Just as toys are considered appropriate for one sex or the other, "few intellectual disciplines are sex-neutral"[11] in our society. Our children are required to mesh their sense of themselves as male or female with their intellectual pursuits and pleasures. By the time children are free to choose course electives, their feelings about the subjects have been heavily weighted.

As we have seen, mechanical and object-related skills become associated with boys as early as toddler play. Later on, math and science themselves come to be defined as antithetical to femininity. Ask a group of people to list characteristics they associate with mathematicians; invite them also to list characteristics they associate with femininity. Chances are they will generate two contrasting lists. Mathematicians are usually envisioned as wise, cautious, rational, impersonal, and asocial. Femininity, on the other hand, is usually associated with emotional, impulsive, irrational, nurturing, and social behavior.

Such stereotypic associations affect teachers and, ultimately, children's expectations of themselves and one another.

---

[11]Sharon Churnin Nash, "Sex Role as a Mediator of Intellectual Functioning," in *Sex-Related Differences in Cognitive Functioning,* ed. Michele Andrisin Wittig and Anne C. Peterson (New York: Academic Press, 1979), p. 291.

Teachers may be even more influential than counselors in determining whether children go on in math and science. Yet many women who teach feel indifferent toward, or even actively dislike, math and science. Many teachers also believe boys and girls have different aptitudes for these subjects. In a survey of elementary and high school teachers, 41% thought boys do better at math and science. Only one teacher thought girls do better at science, and none thought girls do better at math.[12]

Children's peer attitudes follow suit. In third grade, both boys and girls think their own sex superior in math. But by secondary school, both sexes have come to believe that boys are better math students.[13]

Because childhood is the time when gender identity is developing, children are especially vulnerable to sex-role labeling. When whole areas of cognitive functioning, like reading or mathematical reasoning, are labeling feminine or masculine, children's interest and achievement in those areas are jeopardized. Researchers have correlated youngsters' sex-role attitudes with motivation, anticipation of success, and the value they assign to success in subjects. Some of the findings[14] illustrate how powerfully sex-role definitions affect learning:

> Labeling a traditionally masculine task as "feminine" or "neutral" causes sixth-grade girls to raise their expectations of success and to place a higher value on success at the task.

> Six- to eight-year-old girls perform at a higher level when a game is labeled "for girls." Boys do better when the same game is labeled "for boys."

> Females have more confidence than boys on verbal tests, less on spatial tests, and less in "masculine" subjects like math and science.

> Eleven- to eighteen-year-old females place higher value on their performance in verbal, social, and artistic endeavors, than in athletics, natural sciences, and mechanical skills. The oldest girls place the lowest value on math.

> Eight- or nine-year-old boys who see reading as masculine tend to be better readers than boys who think reading is feminine.

> The more feminine a girl rates herself, the higher will be her reading performance; the more masculine a boy's self-image, the greater will be his math performance.

---

[12]John Ernest, *Mathematics and Sex* (Santa Barbara, California: University of California, 1976).

[13]*Ibid.*

[14]For all references on the effect of sex-role labeling on task performance, see Nash, *op. cit.,* pp. 265–68.

The children who are found to be maximally adaptive, that is, those who do well in any cognitive area, are the ones who consider themselves high in *both* masculinity and femininity. Thus, it appears that children reach their fullest potential when they can draw on the cognitive encouragement given both sexes.

SEX STEREOTYPES IN BOOKS

Since relatively few women have traditionally entered math and science fields, girls see few feminine role models to incline them in those directions. In the books we give our children at home and school, we have the opportunity to provide them with examples of men and women developing skills and working at the widest possible variety of jobs. But what do our current school books actually do? In spite of women's studies courses and curriculum materials that have been developed in the last decade, surprisingly little has been mainstreamed into our public school texts.

A recent National Institute of Education review of research on school books concludes:

> ... most current readers and other educational materials not only reflect the society in terms of sexism but even exaggerate reality by portraying society as being more sexist than it is.[15]

Generally, our children still read books that present a distorted profile of the competencies and roles of men and women:

> Women are mainly depicted as housewives and mothers, not as wage earners; men are depicted as sole breadwinners, not as homemakers or child rearers. With more women working outside the home than ever before, including nearly half of all women with preschool children, this portrayal is atypical of American families today.

> When women are depicted as wage earners, they are usually doing traditionally female work. They are involved in nurturing activities, like nursing and teaching, or in providing support services, like secretarial work and waitressing. Few are shown directing projects or businesses, supervising staff, engaging in technical or mechanical work, or conducting research.

> Stories that have women as central characters or concern female

---

[15]National Institute of Education, Constantina Safilios-Rothschild, *Sex Role Socialization and Sex Discrimination: A Synthesis and Critique of the Literature* (Washington, D.C.: U.S. GPO, 1979), p. 73.

achievement are the exception in children's readers. Women are depicted as followers, not leaders. Often shown as inept in problem solving, they rely on men for rescue.

In math and science books surveyed, two out of three pictures were of males. Female pictures sometimes portrayed women as confounded by the simplest computational problem; they asked their children for help or insisted that they wait for daddy to assist them. In word problems, women earned less money than men; boys invented things, while girls used the inventions.[16]

Women are almost entirely absent from history books except, in recent years, for a few pages on women's suffrage or temperance. Both outstanding achievements of women and the activities of the majority of women are omitted. Women's labor in the household, on farms, and in the urban work force is invisible in most of our school books.

Most readers and texts still employ the male pronoun in a universal sense.

Counseling materials and tests usually reflect traditional assumptions about careers for boys and girls and about what course preparation each sex will require for adult life.

Usually, youngsters are not aware of these biases and omissions in their texts. It takes training to recognize them. Children can become so used to viewing the world through male examples that often the inclusion of women, rather than their absence, is the anomaly that they notice.

Children are not the only ones who are affected by images and expectations in textbooks. Adults preparing to teach rely on books to help them decide what and how to teach their students. Teacher-education texts guide prospective teachers in prioritizing issues in learning and in developing classroom methods. But examination of current widely used teacher-education texts reveals that few pay any attention to sex-role stereotyping or its effect on math and science learning. Of the teacher-education texts examined in a study by Myra and David Sadker, 95% gave the issue of sex equity less than 1% of book space. Two out of three science methods books did not discuss the disparity between girls' and boys' science achievement, nor did *any* of the math methods books. Yet all of the reading methods books drew attention to boys' greater difficulty in reading.[17] Thus, teachers receive little guidance in promoting

---

[16]Tobias, *op. cit.,* pp. 83–90.
[17]U.S. Department of Health, Education and Welfare, Women's Educational Equity Act Program, Myra Pollack Sadker and David Miller Sadker, *Beyond Pictures and Pronouns: Sexism in Teacher Education Textbooks* (Newton, Massachusetts: Educational Development Center, 1979).

sex equity. What teachers discover about sex stereotyping and schooling, and what remedies they devise, is largely the result of their sensitivity to learning problems and their own creativity.

In the long run, the content of books and materials contributes profoundly to sex-role labeling. The absence of women in math, science, technical, and leadership roles, and the misrepresentation of female abilities have an impact on children as they formulate their interests and goals. Books do more than articulate subject matter. They are instruments of cultural modeling. They both reflect and help to create our cultural ideas about the kinds of people we are, the possibilities we have, and the visions we cherish. Through narratives and pictures, we extend ourselves back in time and project ourselves into the future.

How might books contribute to girls' avoidance of math and science in particular? Even readers and social studies books tell girls a great deal about their relationship to math and science. Currently, they inform girls that they are essentially passive consumers of technological change. The invisibility of women implies that they have had little impact on the development of science and technology. Furthermore, students are not helped to understand the impact of scientific change on women's daily lives; on the contrary, women's functions as child rearers and homemakers appear immune to change through the centuries. Nor do we read of women's own reactions to the massive technological shifts that have transformed their lives. Why then should girls care about science or believe in their power to affect its course?

In reality, the reasons are profound and plentiful. Science has transformed every aspect of women's lives: housework, birth control, health, the work we do, the foods we eat, the beauty of our environment. To control the developments that so determine our well-being, women must understand and influence the direction of scientific research and decision making. Teachers and parents can choose books and create new materials and examples with sex equity in mind. In so doing, they enable girls to recognize their role in technological and economic development and to begin to pattern a new future.

In this chapter we have seen how lack of practice can adversely affect math and science learning for girls. The toys and games preferred for females fail to develop spatial skills and basic math and science concepts. The labeling of math and science as masculine inhibits girls' interest, motivation, participation, and success. Books and materials often model sexist values and present an erroneous view of women's impact on science and technology and of the impact of science on women's lives. All these factors deter girls from developing essential skills.

Improving skills and promoting familiarity with concepts are major goals of sex equity in math and science. Since manipulating concrete objects is a critical step in concept development, and since such activity is very limited in girls' play, we can help girls by using hands-on activities to teach math and science. Hands-on approaches, or manipulatives, are described in chapter four and used throughout the activities in Part II of this book. Spatial visualization, a particularly troublesome area for girls, is the subject of chapter five. Here, manipulatives and confidence-building strategies maximize opportunities to increase spatial skills.

By modeling diverse roles for men and women, we can alter the messages of sex-biased reading and math books. Sex-role awareness strategies in chapter four describe how math and science materials can be used creatively to reflect women's experience and promote a more accurate picture of women's work and work alternatives.

3 Adolescence and Beyond

In a suburban Syracuse, New York, high school, Francis-Dee Burlin asked 149 eleventh-grade girls to name their ideal career choice and then to name their real intended careers.[1] High school junior Jean explained why she intended to become an airline stewardess, even though airline pilot was her ideal choice:

> Well, I don't know why it is . . . there aren't many girls that are airline pilots. They'd probably be as good as boys, but I don't know if they even wanna do that.

Carol, ideally a lawyer but planning to be a housewife, explained:

> As far as going to school, I can go to about any school I want to, but the thing is I don't know if I am going to go out and become a lawyer after law school. It's like if I become a lawyer, it's a full-time operation and maybe I'll have a husband and I'll want to

---

[1]Francis-Dee Burlin, "Sex Role Stereotyping: Occupational Aspirations of Female High School Students," *School Counselor,* 24 (1976), 102–9.

have kids. I would want to spend time with them and I don't know if I can be a full-time lawyer and a full-time lover, and I don't want to take anything away from my kids or take anything away from my occupation. It bothers me so much. . . . It's just that—I don't know—I want to be somebody that people write down in history, somebody that is not forgotten. I don't want to be just another skeleton in the ground. I think I will become a housewife and it bothers me because I don't know what to do.

Judy told the interviewer why her plans diverged from the ideal:

When a person has a boyfriend you have to act inferior when you really aren't at times, because, you know, you have to be careful. So if you want to have a boyfriend or husband or anything, you have to feel a little inferior.[2]

Although the ideals of these girls and the majority of their classmates reflect new horizons for women, their plans were modified by traditional sex-role expectations. On a questionnaire, most girls said it was training or talent or money that kept

FIGURE 3-1. *Photo by Richard Simpson.*

them from pursuing ideal choices. But their later interviews revealed another tale: conflict between careers and family; reining in talent and intelligence to gain social acceptance; struggling with values they still hold but increasingly experience as restricting.

---

[2]Student quotes from Burlin, *loc. cit.,* pp. 104–6.

Are Jean, Carol, and Judy unusual in their sentiments? Or do they share the doubts and conflicts of a generation of young women? After a decade of the women's movement, their words have a familiar historical ring and seem strangely discordant with the much publicized changing life styles and work patterns of women. After all, many adolescent girls today readily proclaim, "I can be anything I want to be. Women no longer face the restrictions of our mothers' generation. Today women are doing everything!" To hear these youngsters, and to be acquainted with women who are electricians and physicians, leads us to believe that occupational barriers based on sex no longer exist. All too often adults cannot understand why girls don't take advantage of these apparent new opportunities.

Equality for women is a principle that has been on the ascendency during the last decade. But the principle still eludes most people in the daily practice of their lives. Today, more youngsters than ever before have grown up with working mothers and accept the notion of women working outside the home. But girls today are in a triple bind: the traditional vision of the full-time wife and mother conflicts with economic realities which dictate that most women must work; the newer vision of the career woman conflicts with child rearing, which is still seen mainly as women's responsibility; and the new ethic of equality inspires girls to be more independent and competent than they suspect is really acceptable and more than they have been taught to feel. To reconcile old and new demands, and their own feelings, girls would have to become superwomen.

Since adolescence is the time when we first come to terms with adult roles, it is then that the clash between principles and realities, between achievement and social fulfillment, makes its first major impact on the lives of young women. What roles math and science will play in young women's plans depends on how they resolve sex-role pressures in the adolescent years.

As most of us remember, adolescence is an especially vulnerable time—a time of sweeping physical and emotional changes. In this passage out of childhood, the need for social approval escalates, as does the pressure to conform. Boys and girls are expected to identify with the "appropriate" sex role as they carve out niches in an unknown adult world. Adolescents are moving away from parents, as yet unsure of who or what will take their place, and are testing out new roles with the opposite sex. The culture of adolescence is sometimes brutal, with social judgments made quickly and often on superficial grounds, placing friendships and status at risk.

When their children reach adolescence, parents also feel anxious about separation and sex roles. They are often unsure whether their children will be mature enough to handle the changes in their lives. They have great hopes for their children and wish to protect them from mistakes that will affect their futures, particularly social mistakes for daughters. The guidance they provide is filtered through their feelings about the decisions they made and pressures they felt in their own youth.

At school, adolescence is the time when the "feminine" world of the elementary school is superceded by the increasingly "masculinized" world of the junior and senior high. The school environment becomes less intimate, and more and more principals and teachers, particularly math, science, and shop teachers, are male. Increasingly, too, what happens in school is linked to future careers. As courses become more advanced throughout the high school years, math and science classrooms become more male, with fewer girls enrolling. A girl who wishes to pursue more advanced work finds her fear that "girls don't become scientists" reinforced daily by the ratio of boys to girls in the classroom. To succeed, she must compete directly with the very males from whom she is trying to win social acceptance and about whom Judy said "you have to be careful."

Given the burdens and insecurities of the adolescent years, girls need particularly strong support and enlightened guidance to pursue subjects like math and science. Instead, in the ninth grade many teenagers are presented with their first option to drop these subjects. Seeking help from parents and counselors in making course decisions, girls are often influenced by misinformation about the impact of these decisions on later options and by sex-biased assumptions. Even very egalitarian-minded counselors report reacting differently to a girl who wishes to drop back into a less advanced math class than to a boy, who "is more apt to need the math." Likewise, concerned parents are sometimes happy to relieve a daughter of a worrisome chemistry class, but in similar circumstances encourage a son to stay in the course for the sake of his future.

These responses exacerbate the intellectual self-doubt girls carry with them from childhood. Although a C in calculus or chemistry may be a greater credit to a student and a greater help to her future than an A in another course, many girls construe a C, or even a B, as failure. Adults who allow these girls to drop the course without questioning their sense of failure collude in the youngsters' feelings of helplessness and their limiting of future options.

Since adolescents are also strongly influenced by peer culture, it is important to understand what young women believe their peers will think of their achievement and how this affects their behavior. Do boys really find mathematically or scientifically talented women less attractive, or is this merely a fear girls have? Studies of young people in the sixties and seventies show that although boys may be punishing girls less for being smart, they still harbor negative attitudes and that those attitudes still affect girls' behavior.[3] Among the results, we find:

adolescents think less of girls who are good in mathematics than of boys who are good in mathematics;

boys are especially likely to characterize mathematics as a male domain;

not surprisingly, gifted girls are hesitant to accelerate in mathematics because they fear their social life will suffer;

among the highest-ranking reasons for girls' disinterest in mathematics is "males do not want females in the mathematical occupations."

It is evident that no matter what course decisions a young woman makes, these peer judgments cannot entirely escape her awareness. Chances are they affect her feelings about herself, feelings she must come to terms with as she works and plans.

Considering peer pressure, adult expectations, and the tentative independence of girls, we might well imagine that academic success in "male" fields is a loaded issue for a girl. Following the work of Dr. Matina Horner,[4] many researchers have even argued that women may stifle their own ambitions because of a "fear of success." According to this theory, women are more disposed to become anxious about achievement because they fear negative consequences. Horner found that college women responded to stories or pictures involving the success of a woman by fantasizing rejection or bad feelings or by denying that the woman was responsible for the achievement. Interestingly, success anxiety was especially acute in talented women for whom success was indeed a real possibility. It was also high in mixed-sex competitions where a woman might be torn between her competence and her desire for male approval.

---

[3]For a review of the literature on attitudes towards women in mathematics, see Sharon Churnin Nash, "Sex Role as a Mediator of Intellectual Functioning," and Lynn H. Fox, Diane Tobin, and Linda Brody, "Sex Role Socialization and Achievement in Mathematics," in *Sex-Related Differences in Cognitive Functioning,* ed. Michele Andrisin Wittig and Anne C. Peterson (New York: Academic Press, 1979), pp. 263–332. See also Lynn H. Fox, "The Effects of Sex Role Socialization on Mathematics Participation and Achievement," in NIE Papers on Education and Work No. 8, *Women and Mathematics: Research Perspectives for Change* (Washington, D.C.: U.S. GPO, 1977).

[4]Matina S. Horner, "Toward an Understanding of Achievement-Related Conflicts in Women," in *And Jill Came Tumbling After: Sexism in American Education,* ed. Judith Stacey, Susan Bereaud, and Joan Daniels (New York: Dell Publishing Co., Inc., 1974), pp. 43–63. For a review of the literature on achievement motivation and women, see National Institute of Education, Constantina Safilios-Rothschild, *Sex Role Socialization and Sex Discrimination: A Synthesis and Critique of the Literature* (Washington, D.C.: U.S. GPO, 1979), pp. 27–42.

In an informal variation of one Horner experiment, junior high school students were asked to react to the success of Susan and John, fictitious students who find themselves ranked highest in their medical school class after first-term finals. The responses of the top two students in a junior high algebra class, one a boy and one a girl, reveal strong gender differences in success anxiety.[5]

Tim: I think John's reaction would probably be one of satisfaction. If he could make it all the way through college and now get high grades in first year med school, he should be proud and be happy to continue medical-related work as he is probably suited for it.

Cindy: Most people probably would feel that Susan would be proud and really happy, but I don't feel that way. Susan will have to keep up the reputation of being the person at the top of the class. This could put a lot of extra work on her. The teacher will expect more of her than of everyone else. Being the top student is hard; it's so much easier to be the second or third student from the top. The teacher might also think that Susan isn't working up to her ability if she doesn't remain at the top. Susan might also worry about how the other students will feel about her. Could they resent her? The smarter students could be jealous because Susan seems to be smarter than them. This jealousy could make them create a competition between themselves and Susan. I don't think Susan will feel the pride and the happiness too long.

We should not conclude from Cindy's words that she wants success any less than Tim. She is, in fact, a top student. Unlike Tim, however, she is highly sensitive to the interpersonal problems accompanying success and is uncertain about her ability to live up to her reputation. Both feelings are products of gender socialization and real bias against women's achievement. These feelings are evoked especially when the achievement is in a traditionally male field. It is hardly likely, after all, that women would fear becoming good cooks or caring mothers. Understandably, they may fear what they are taught they "ought not to have." While for boys math and science successes can heighten masculine self-esteem, girls must walk a psychological tightrope between pride in their achievement on the one hand and a threat to feminine self-image and social support on the other.

The fact that many young women do become good mathematics and science students in spite of anxieties is a

[5]Data collected by the Math/Science Sex Desegregation Project, Novato (California) Unified School District, 1980, Joan Skolnick, Director. Project funded by the U.S. Office of Education. Student names fictitious.

tribute to their fortitude and intellectual self-respect. Yet even among high school girls who do exceedingly well in the most advanced courses, such as chemistry, physics, and calculus, a great many never consider those fields as majors or careers. According to national studies, a major reason for this is that girls do not see mathematics as relevant to their future goals.[6] This view is confirmed by what high-achieving high school girls themselves observe about their male classmates: "Boys have clearer career goals and a clearer understanding of how math and science courses will help them reach these goals."[7]

Why is it different for girls? We cannot attribute the difference to lack of interest, since girls who have elected to take physics and calculus are presumably interested. We do know that girls lack exposure to a great range of careers that utilize math and science, need better career education regarding course and skill requirements, and need more female role models.

Still, there is a more fundamental problem. After they leave school, most women will live and work in dual worlds: the world of the family and the world of the labor market. Some girls still envision work and family as mutually exclusive choices. Others enthusiastically envision doing both, provided they can find a mate with whom this is possible. Some are keenly aware that work is a necessity, not an option. All girls, however, face a potential conflict. Boys also want families and see them as essential to their future happiness. But far from learning that career jeopardizes family life, boys are led to believe the opposite: that job success is the way to win love and maintain family life. A boy who is career-oriented, and plans his courses accordingly, is propelled forward through a green light by the momentum of social pressure. For girls, the light is always yellow, and the sign reads, "Caution: Proceed Only When Road Is Clear of Family Responsibilities."

Making room for future family commitments tends to orient girls toward *jobs* rather than *careers,* that is, toward occupations which they imagine involve less training and less single-minded devotion to professional development. Stereotypically, math and science occupations are thought to fit the career mold, perhaps precisely because they have traditionally been male domains. Although it sways girls away from math and science fields, the notion that traditionally female occupations are more conducive to family life is questionable. Many times these occupations are merely lower-paying. English majors who

---

[6]Lynn H. Fox, Elizabeth Fennema, and Julia Sherman, *Women and Mathematics;* Fox, Tobin, and Brody, *loc. cit.,* pp. 312–14.

[7]Based on data collected by the Math/Science Sex Desegregation Project, Novato (California) Unified School District, 1980.

become secretaries spend the same amount of time in school as engineers with Bachelor of Science degrees and more time than some electronics technicians. In addition, their jobs are usually no more flexible than the jobs of their male counterparts. In Europe, dentistry is considered a feminine occupation because it is "flexible and compatible with motherhood."[8]

FIGURE 3-2. Printer.
Photo by Richard Simpson.

So long as women are primarily in charge of child rearing, career planning for girls can never be fully separated from reference to family, as it has been for boys. What is relevant to a girl's future will be calculated on the ebb and flow of family needs, as well as on the merits of particular careers. To address this problem properly, good career education must present new perspectives on nontraditional careers, as well as help both boys and girls envision a new shared role within the family.

How have the views of young men and women on family responsibilities been changing? The evidence suggests that although more women plan to work continuously, the majority of young women in high school and college still define marriage and motherhood as primary goals, still assume *they* will care for the children, and still tend to give the man's career priority.[9]

---

[8]Safilios-Rothschild, *op. cit.,* p. 43.

[9]Helen S. Astin, "Overview of the Findings," in *Women: A Bibliography on Their Education and Careers,* ed. H. Astin, H. Suniewick, and S. Dweck (New York: Behavioral Publications, Inc., 1974), pp. 1–10. For a review of the literature on attitudes towards women's careers and family responsibilities, see Safilios-Rothschild, *op. cit.,* pp. 44–46.

Often this is true even among women who are remarkably free of other kinds of sex-stereotyped ideas. In Burlin's study of the ideal and real career choices of high school girls, all but one girl interviewed wanted children, and three fourths of them said they would stop working when they had children. "None of those interviewed seemed to have considered the possibility of an egalitarian marriage, which provides the framework for both husband and wife to pursue careers, and at the same time, participate equally in the pleasures and responsibilities of raising children."[10] Apparently, the idea that men should work within the home has been slower in gaining acceptance than the idea that women should work outside.

Young women's estimations of what is expected of them reflect certain realities. Young men appear to be even more traditional in their perspective on family roles.[11] Though most high school boys today grew up with mothers who work outside the home, a classroom discussion of equal opportunity for women can still generate a vociferous defense of manhood and conjure up images of women deserting home and family.[12] Interestingly, among college men, it is math, science, and business majors who insist on the least career involvement for their wives.[13]

How do people reconcile these attitudes with the increasing economic necessity for women to work? Media images and today's adult generation give youngsters some models that are problematic. In a *Ms.* magazine article entitled "She Brings Home the Bacon and Cooks It Too: Madison Avenue Thinks They'll Keep Her," Bernice Kanner summed up the results of two hundred surveys among eighteen- to fifty-year-old men:

> Men seem to be all for contemporary women as life partners—as long as they pursue their outside interests, including work for pay, after attending to household chores they've traditionally handled.[14]

Many women are trying to fulfill these expectations and feel guilty when they do not succeed. The image of women "whipping up the beef wellington after a hard day at the nuclear physics lab"[15] is tantalizing. But it is a fantasy that most girls

---

[10]Francis-Dee Burlin, *loc. cit.,* p. 104.

[11]Safilios-Rothschild, *loc. cit.*

[12]Based on discussions with secondary school teachers and students in the San Francisco Bay area.

[13]M. R. McMillin, "Attitudes of College Men Toward Career Involvement of Married Women," *Vocational Guidance Quarterly,* 21 (1972), 8–11.

[14]*Ms.,* March 1980, p. 104.

[15]Lindsey Van Gelder, "The Selling of the 25-Hour Day, or A Case of Organization Guilt," *Ms.,* March 1980, p. 47.

know they cannot live up to. It provides little incentive for girls to attempt demanding careers in math and science because the image is a setup for failure. We as parents and educators can help girls and boys face their futures realistically so that they can develop new solutions to the dilemmas that await them. This support will free young women to consider math and science careers as relevant and feasible in their life plans.

This chapter has discussed the conflicts and pressures that discourage adolescent girls from continuing in math and science. These include conflict between work and family, heightened pressure to conform to sex roles during adolescence, negative attitudes of peers about girls who do well in math and science, expectations and biases of parents and counselors, "fear of success" in male fields, poor career education, and few role models of women teaching math and science and working in related professions.

How can we create more positive influences so that our children will make the best course and career decisions during adolescence and beyond? The remainder of this book presents strategies and activities that address this question. Sex-role awareness strategies and social arrangements strategies, described in chapter four and used throughout our activities, are particularly designed to assist children with the social problems raised in this chapter. With heightened sensitivity to peer pressures, social arrangements strategies help to create more socially comfortable learning situations. Sex-role awareness strategies include utilizing math and science problems as career education tools and involving parents and teachers as role models. These strategies acknowledge the sex-role pressures with which our children must cope, and support their pursuit of their interests, regardless of sex.

Life for the next generation of adult women and men will be different. As families adjust to the working lives of both adults, pressure to redefine roles within the family will increase. Our children feel the crosscurrents of these changes already, and we cannot afford to ignore them in our teaching, parenting, and counseling. Whether children defend traditional roles or proclaim liberation, chances are they are feeling some of both, and most are eager to talk about sex roles. Since we did not reach adulthood in the same environment, we cannot tell them exactly what's ahead. But we can help them develop the personal strengths, intellectual tools, and social supports to plan for it.

4

Strategies to Develop Math and Science Skills

In the preceding chapters, we outlined ways in which gender socialization negatively affects girls' attitudes towards math and science and inhibits the development of their skills to full potential. Female socialization promotes fears about competence and independence, reliance on the judgment of others, an interpersonal and verbal orientation, and unfamiliarity with toys, games, and activities that stimulate learning of spatial relationships and basic mathematical concepts.

What teaching strategies, then, can we use to assist girls in overcoming early influences? Below, we identify and discuss several key strategies. Part II of this book (chapters five through eight) is devoted to math and science activities that utilize these strategies in a multitude of ways.

Problem solving is at the center of our approach. Our strategies help children build competence and confidence in problem solving, and the activities in Part II exemplify the diversity of problem-solving tasks in math and science. Organizing mathematics and science learning around problem solving is now widely advocated by mathematicians, scientists, and educators. In *An Agenda for Action,* the National Council of Teachers of Mathematics makes the following recommendation: "The

FIGURE 4-1. Paper Airplanes, Chapter 8. Twin Pines School, Oakland, California.
Photo by Richard Simpson.

development of problem-solving ability should direct the efforts of mathematics educators through the next decade. Performance in problem solving will measure the effectiveness of our personal and national possession of mathematical competence."[1]

Our strategies fall into four main categories, each of which is the basis for a section of this chapter. Strategies in the first section focus on building confidence and reducing fear of taking risks. The second section discusses the value of manipulative materials, that is, an activities approach to math and science learning. The third section, concerned with the social arrangements of activities, describes how independent work, group work, cooperation, and competition enhance learning. Strategies in the final section help to build sex-role awareness, without which children will not question stereotypic attitudes, and many girls will neither understand how studying math and

[1]*An Agenda for Action: Recommendations for School Mathematics of the 1980s* (Reston, Virginia: The National Council of Teachers of Mathematics, Inc., 1980), p. 2.

science benefits them nor develop the skills necessary to succeed in these fields.

BUILDING CONFIDENCE

Confidence is inextricably linked to problem-solving skill: expert problem solvers are people who believe they can do it. Due to early socialization, girls are more likely than boys to believe that they can't do it and are therefore more likely to get discouraged and give up when confronted with a difficult problem. As a result, they lose critical practice and tend to fall behind boys in problem-solving competence. This further reduces their confidence. The following confidence-building strategies will help counteract female "learned helplessness" and disrupt this insidious cycle:

Success for each child
Tasks with many approaches
Tasks with many right answers
Guessing and testing
Estimating

Success for each child means that every child has the opportunity to be successful at her own level. Many different kinds of activities afford this opportunity. For example, children doing the same science experiment may have many different levels of understanding, and children playing the same strategy game may have many different levels of skill. *Tasks with many approaches* and *tasks with many right answers* are especially well suited to enabling children to be successful at their own levels. In our activity Making Shadows, children are asked to identify pairs of objects that could make similar shadows. Some children will identify only one or two pairs, whereas others will readily identify many. Almost all children will have some level of success at this task and thereby improve their spatial skills.

Tasks with many approaches and tasks with many right answers are also valuable because they enable us to reward many different kinds of problem-solving behavior, such as working backwards, organizing information into charts, or drawing a picture of the problem. This is essential if we are to help children develop versatility in problem solving. In both math and science, the process is just as important as the answers. To give an example from science, asking questions is an important skill in scientific investigation. This skill is emphasized in our activities Asking About Animals and Patterns in Nature. If we reward only answers, we are downplaying process and giving children an unrealistic perspective of math and science.

49

FIGURE 4-2. Paper Airplanes, Chapter 8. Twin Pines School, Oakland, California.
Photo by Richard Simpson.

Guessing and testing is both a valuable teaching strategy and an important problem-solving strategy used by mathematicians and scientists. We guess approaches to problems as well as solutions to problems. In science, guesses are called hypotheses. In both math and science, it is necessary to take risks. A mathematician or scientist who routinely refused to carry out procedures that might not lead to correct answers would never accomplish anything. Mathematicians and scientists know that wrong answers are valuable: they lead to refined guesses and bring us that much closer to the solution.

Distrusting their capabilities, girls are often afraid to guess—they are afraid to be wrong. They may even feel that guessing is inappropriate. Activities in which guessing is encouraged help to develop confidence in risk taking as well as skill in hypothesis formulation and assessment. In many of our activities, such as Card Scramble and Exploring Symmetry, only the child knows how she guessed. Without fear of being judged, the child can discover that guessing can, in fact, lead to solutions.

Estimating, then comparing the estimated answer to the exact answer, helps build confidence in mathematical ability. Children are surprised by how close they can come to the exact answer. Like guessing and testing, estimating is not only a valuable teaching strategy but also an important skill in itself.

Many problems in math and science do not require an exact answer; often, an estimation is more useful.

MANIPULATIVE MATERIALS

Most of the activities in this book include the strategy:

Using manipulatives

We suggest a variety of easy-to-find materials to provide hands-on experiences with math and science concepts. For example, we use materials such as playing cards, sugar cubes, and boxes in the math activities and clay, mirrors, rocks, and leaves in the science activities.

As we have noted, because of socialization—early childhood practices, mothering, toys, and games—a girl is encouraged from the earliest years to learn through close verbal interaction with others rather than through the more male approach of independent manipulation of objects in the environment. She then moves on to schools in which traditional

FIGURE 4-3. Hundreds Puzzle, Chapter 6. Twin Pines School, Oakland, California.
Photo by Richard Simpson.

paper-and-pencil models of teaching require that she interact with science and math concepts on a purely verbal, written, or symbolic level without first experimenting with the concrete materials. What has happened here? Does socialization predispose girls to difficulty in mathematical or scientific problem solving by pushing them to skip steps in the learning process?

Piagetian scholars believe that before a child acquires the ability to manipulate and abstract concepts mentally, she or he must pass through a stage of concretely manipulating objects. "It is the experience with the materials of the discipline that produces the person who can understand abstract content . . ."[2] For example, by designing and performing experiments with concrete objects, a child ultimately acquires the ability to design controlled experiments in his or her head. This process occurs with many kinds of scientific and mathematical reasoning. The stage of interacting with objects is called the concrete stage; the stage of abstract reasoning is called the formal operational stage. Piagetians have concluded that no matter how old the student is, no verbal or written instruction can replace what must be learned through physical manipulation of objects.

The educational message here is that opportunities for independent, self-regulated, concrete learning are necessary to promote full intellectual development. As we have seen, opportunities for mechanical and visual play, which build important math and science skills, are likely to be restricted in female socialization. Although boys as well as girls receive primarily paper-and-pencil instruction, boys are more likely to gain in play what they have missed in school.

Piagetian scholarship argues persuasively that it is critical to use manipulatives in teaching math and science. Hands-on activities afford girls missed opportunities to experiment with concepts such as size, shape, distance, and speed. In doing so, they begin to visualize and conceptualize relationships that may be second nature to boys, who have had prior experience with their use and importance. Exploratory and open-ended involvement with concrete objects may then be followed by work with abstract representations, such as equations and graphs.

Manipulatives should be used not only at the elementary level but also at the secondary level. Many adolescents and even adults find science and math difficult because they lack the concrete experience from which to make sense of concepts.

---

[2]Anton E. Lawson and John W. Renner, "Piagetian Theory and Biology Teaching," *The American Biology Teacher* 37 (1975).

SOCIAL ARRANGEMENTS

How and what children learn is affected by what happens in social groups—in families, in peer groups, in classes. Unfortunately, our social arrangements sometimes negatively affect girls' learning in math and science. We can begin to improve the learning environment by paying attention to the social interactions we set in motion while children are at work.

FIGURE 4-4. Make a Picture, Chapter 5. Twin Pines School, Oakland, California. *Photo by Richard Simpson.*

Every learning task involves a social arrangement. Children may work alone, in pairs, in groups, or in single-sex or mixed-sex groups. Activities may be structured to maximize cooperation, competition, or a healthy combination of both; competition may be set not only between individuals but also between groups, thereby creating cooperation within a competitive framework. Independent work can be done in a supportive, nonpressured atmosphere or in an atmosphere in which speed and results are stressed. We refer to all these possibilities as the social arrangements of activities. Sensitivity to social arrangements will enable us to help girls in math and science. The strategies in this section emerge from the various ways children can be grouped for mathematics and science learning:

Independent work
Cooperative work
Single-sex and mixed-sex groups
Game playing

Children gain confidence in their mathematical ability when allowed to perform *independent work* at their own rate, and this confidence is particularly important for girls. Too often, there is pressure to work quickly when doing mathematics. There is a common misconception that success in math is based on speed. In reality, faster does not necessarily mean better, and problem solving takes a lot of time. It is often helpful to examine a problem, leave it for a while, think about it, and return to it at another time with a fresh perspective. At home and in the classroom, we can structure problems that children can think about and work on over a number of days. This eliminates the pressure to find a solution in one sitting and emphasizes the importance of the child's own thinking process.

Girls and boys grow confident and able by building on what they have learned to do easily, well, and with interest. From a young age, female socialization emphasizes social skills, a more cooperative style of learning, verbal skills, and a keen interest in relationships and the daily lives of people. If we structure our math and science tasks around these competencies, girls will find their involvement less anxiety-ridden, more interesting, and more natural in conjunction with their other activities. *Cooperative work* will maximize opportunities for talking and interaction among children. Boys will be encouraged to build competency in social and verbal skills, which tend to be more problematic for them. At the same time, children will see each other in less stereotypic terms. Girls will see boys sharing, talking, and questioning. Boys will see girls as valuable partners in solving math and science problems.

In exchanging information and discussing problems, children do not feel the isolation that is often connected with intellectual work, especially in the fields of mathematics and science. Children often feel more comfortable asking questions of their peers than of an adult. By asking questions and helping others, real teaching and learning take place. Comparing information and answers and discussing how to arrive at solutions are valuable experiences for children. A child can increase her own understanding of a problem by explaining how she solved it.

In structuring activities we should be sensitive to peer-group pressures. We have seen that, especially in adolescence, girls are influenced by the negative attitudes of boys towards "math whizzes." As a result, some girls become anxious about succeeding, particularly when they are competing with boys.

One strategy to try, then, is setting up activities so that girls have some time to work with one another in *single-sex groups*. Single-sex groups are also important because girls have not had the informal math and science experiences boys have had and may be intimidated in mixed-sex competitions. *Mixed-sex groups* can be postponed until the later stages of an activity, after girls have had a chance to gain confidence and support from their female peers. The activity Paper Airplanes illustrates using single-sex groups at the initial stages of an activity and mixed-sex groups later on. When boys and girls work together cooperatively so that their group can compete successfully, a female "math whiz" becomes a cherished member of the team.

Game playing involves another grouping that can enhance children's learning. In a game, the moves of one player provoke the thinking of the other players. Two-person strategy games are particularly effective in improving problem-solving skills. Children learn to plan their moves by examining many possibilities, to think about the consequences of their actions (what will the other player be able to do if I do this?), to plan ahead more than one step at a time, and to visualize future moves. These are all valuable skills in logical reasoning.

SEX-ROLE AWARENESS

The strategies we have discussed provide innovative opportunities to increase girls' skills in math and science. But in order to take advantage of their learning, and to consider working in math and science fields, girls will need to challenge stereotypes both in themselves and in others. The following sex-role awareness strategies help children develop the tools to confront social barriers and to deal with the social consequences of stepping into new roles:

> Content relevance
> Modeling new options

Although children pick up a great many indications of what they should or should not be because of their sex, they are rarely given the chance to question their feelings about these messages. The older they get, the more solidly internalized is the world of sex roles, the more these messages are accepted as facts of life, and the greater are the risks involved in questioning them. Through *content relevance* and *modeling new options*, adults can provide a safe, structured environment within which to help children clarify their own values, reassure them that it's all right to consider new possibilities, and increase their awareness of how rigidly defined sex roles obstruct opportunities for both sexes.

The content of math and science problems can provide a fundamental kind of career education. It can develop new perspectives on work and on the abilities of both sexes. Word problems can inform children about occupations unfamiliar to them and provide information about the labor market, such as jobs and salaries, women's position in the work force, and math and science courses they will need for particular careers. Illustrating many kinds of content relevance, creative math and science problems encourage girls' pursuit of nontraditional occupations.

In addition, math and science instruction can address the interests and concerns of girls by presenting a more direct

FIGURE 4-5. Surveyor.
Photo by Lyn Reese.

connection between math and science work and the lives of people. Math and science teach us to solve the problems of living—to care for animals, grow food, build houses, cities, and parks. They are essential to activities as diverse as sports and cooking. Yet much of the time, math and science are presented as abstractions, devoid of human content. Children of both sexes will benefit from activities that involve them in solving the human problems mathematicians and scientists solve. See, for example, our math activity, Facts About Food, and our science activity, Heart Rate.

Word problems can also be useful in modeling new options. They can portray women working at nontraditional jobs and using math and science in their daily lives; they can portray men in nurturing roles. They can also teach children to see women's work as valuable and necessary and teach girls to identify work outside the home as a permanent feature of their lives. Children who understand that a woman wage earner is not an anomaly but an American norm can make realistic course decisions and plan effectively for their futures.

You are an important resource for your children. You can model new sex-role attitudes and let children know that it's all right to try new things. The following are examples of adult role-modeling: a male teacher or a father spending time in the doll corner with young children; a female teacher or mother participating in a science experiment; a teacher or parent asking a girl to join her in a building project. These easy measures can make a profound impression and go a long way toward making a task acceptable for boys and girls who are shy about stepping out of sex-role boundaries.

II ACTIVITIES FOR SEX-FAIR LEARNING

The math and science activities in this section will help develop the skills that research and standardized tests have shown to be problematic for girls. These skills are:

1. spatial visualization, including graphing (chapter five),
2. problem solving, applications, and understanding numbers (chapter six),
3. logical reasoning (chapter seven), and
4. scientific investigation (chapter eight).

Although these skills are crucial in higher science and mathematics, traditionally they have not been emphasized in the curriculum until high school. By this time, due to differences in male and female socialization, girls are falling behind boys in these skill areas and are already showing lack of confidence and patterns of math/science avoidance. Based on the strategies outlined in chapter four, our activities will help girls build a solid foundation for later study, develop confidence in themselves as math and science learners, and broaden their gender expectations by envisioning themselves as future workers in a variety of math and science occupations.

Since many of the skills listed above are used in the social sciences as well as in the natural sciences and mathematics, the activities will benefit girls who ultimately go into fields such as psychology, sociology, and economics. Social science research involves both scientific investigation and logical reasoning. Researchers ask questions, formulate hypotheses, and design studies. In planning surveys and analyzing data, they must use logical reasoning (probabilistic reasoning) to select and apply appropriate statistical methods and tests. Social scientists often display their results in tables or graphs.

By constructing models, performing investigations, discovering problem-solving strategies, and visualizing relationships between two- and three-dimensional objects, children will begin to think like scientists and mathematicians. These kinds of tasks, if encouraged at home and incorporated into the curriculum at every level, will contribute to the intellectual development of both boys and girls. Additionally, these tasks will help build positive attitudes. Too often, the fun and excitement of math and science become lost in rote drill and memorization of facts.

One or more math or science concepts are presented at the beginning of each activity. These concepts are for your information; prior familiarity with them is not needed in order to do the activities. In most cases, knowledge of the concepts will result from doing the activity. In the few activities for which a concept is a prerequisite, instructions are given for presenting the concept to the child.

These activities should be considered a beginning in sex-fair math and science instruction. They show how to build the skills, awareness, and confidence that do not currently ensue from girls' experiences at home and school. We hope that you will also use activities from the books and materials on the Resource List and that our activities will serve as prototypes that inspire you to create your own.

5 Spatial Visualization

MATH ACTIVITIES

Spatial visualization skills, generally not emphasized until high school geometry, can be incorporated into the curriculum at every level. They can also easily be encouraged at home. Many kinds of enjoyable activities, such as puzzle solving, model building, strategy games, construction tasks, and drawing utilize spatial visualization.

The skills developed in our mathematics activities on spatial visualization include:

1. recognizing shapes and their relationships to each other;

2. seeing relationships between two-dimensional shapes and three-dimensional objects;

3. recognizing properties of three-dimensional structures;

4. locating points on a plane by using coordinates;

5. using graphs and tables to organize information.

61

These skills cannot easily be developed through traditional paper-and-pencil drill.

Learning to use graphs and tables is especially important, because collecting and organizing information is a basic mathematical skill. Graphs and charts help us to visualize and comprehend information. Statistical information is often presented in graph form. We use visual representations not only in formal mathematics but also in our daily lives.

Encouraging the early development of spatial skills will help prepare children for later study of higher mathematics. Here they will use these skills in a variety of ways. In geometry, they will examine two- and three-dimensional structures and their relationships. In analytical geometry, they will represent abstract mathematical equations on two or more dimensional coordinate graphing systems. Visualizing the graphs of equations in this manner is an aid to problem solving in calculus. Sketching a graph or other picture of a mathematical problem often aids us in reaching the solution.

Making Pentominoes and the Tangram Puzzle are activities in which children have a chance to recognize shapes, move them around, and see the relationships among them. They can also construct new shapes from the puzzle pieces. Hands-on experiences of this type form a basis for visualizing rearrangements and folding of objects. Children learn to examine a figure carefully from different perspectives in Hidden Shapes. They also see how small shapes can be combined to make larger ones. In The Last Piece and Match the Faces, children must rotate objects mentally to see how they fit within particular spaces.

Several activities give children the opportunity to construct three-dimensional objects from patterns in two dimensions and vice versa. In Folding Boxes, children first visualize folding patterns into boxes, then actually do it. They construct solid figures from flattened patterns in Make a Mobile. Milk Carton Cutting involves cutting and flattening milk cartons to form two-dimensional patterns. Match the Faces, which entails matching a box with pictures of its faces, provides yet another way of relating a three-dimensional structure to its two-dimensional components.

Constructing and examining three-dimensional structures is the focus of the activities included in Building with Cubes. Children learn to visualize cubes in structures even though some of the cubes cannot actually be seen. As a result, two-dimensional drawings of three-dimensional structures become more meaningful. It is not easy for children to see the relationships between book drawings and the structures to which they refer.

Several activities help children learn to organize information and construct and interpret graphs. In Find the Squares, a table is an aid to discovering patterns in numbers. Grid Games provide an enjoyable way to practice locating points on a grid. Add-A-Coin, one of the games, uses both graphing skills and computational skills. Making Graphs gives children experience in organizing information and discerning relationships.

Tangram Puzzle

SKILL AREA: Spatial Visualization

GRADE LEVEL: Primary, intermediate, junior high

STRATEGIES: Many approaches
Many right answers
Using manipulatives
Independent work
Cooperative work

MATH CONCEPTS: *Square, triangle, parallelogram,* and *rectangle* are the names of common two-dimensional shapes.
Congruent shapes are the same shape and size: they fit on top of each other.
Similar shapes are the same shape but a different size.
The following geometric terms may also be introduced while making the tangram, depending on the level of the child: right triangle, isosceles triangle, trapezoid, midpoint, base, vertex.

Make Your Own Tangram

MATERIALS: A 6-in. (or 15cm) square cut from construction paper, scissors

DIRECTIONS: This activity gives you directions to follow to make your own tangram puzzle from the square. The next activity describes specific tasks to do using this puzzle. Have the child make a tangram puzzle.

1. Fold, then cut your square in half diagonally.

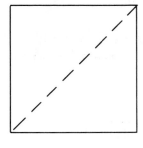

FIGURE 5-1.

2. Fold, then cut one of the triangles in half. Set these two new triangles aside.

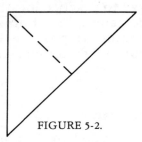

FIGURE 5-2.

3. Take the other large triangle. Find the midpoint of the long side. Fold the triangle so that the corner (vertex) opposite the long side touches the midpoint. Cut and set the triangle aside.

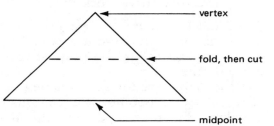

FIGURE 5-3.

4. Fold the trapezoid, then cut it in half.

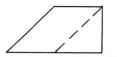

FIGURE 5-4.

5. Take one of these halves and fold as illustrated. Cut, then set aside the two pieces, a square and a triangle.

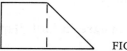

FIGURE 5-5.

6. Take the remaining piece and fold as shown. Cut, then set aside the two pieces, a triangle and a parallelogram.

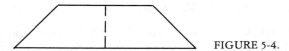

FIGURE 5-6.

7. Now you should have all seven pieces of the tangram puzzle: 2 large triangles, 2 small triangles, 1 middle-sized triangle, 1 square, and 1 parallelogram. In this set the two large triangles are congruent to each other, as are the two small triangles. The large triangles are similar to the small triangles.

Using Your Tangram Puzzle

MATERIALS: The set of 7 tangram pieces, construction paper, pencils

DIRECTIONS: Have the child work with a friend. Each child makes a design or shape with all seven pieces. This is done by placing the pieces on construction paper and tracing around the outside of the shape. Have children exchange papers and try to fill in each other's outline with their tangram pieces.

Challenge the children to try the following: make many different-sized squares; make many different-sized triangles; make many different-sized rectangles; make the first or last initial of their names; make a large square using all 7 pieces.

Hidden Shapes

Both of these activities give children an opportunity to examine a picture carefully to find the different ways that shapes fit together to make new shapes.

Find the Triangles

SKILL AREA: Spatial Visualization

GRADE LEVEL: Primary, intermediate

STRATEGIES: Success for each child
Many right answers
Many approaches
Independent work

MATH CONCEPTS: Small triangles can fit together to make larger triangles.

Triangles can fit together to make different shapes such as *diamonds, trapezoids,* and *hexagons.*

MATERIALS: Several copies of the diagram made up of triangles shown in Figure 5-7.

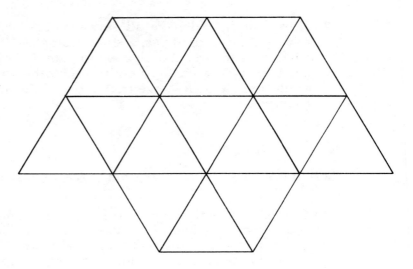

FIGURE 5-7.

DIRECTIONS: The child is to try and find as many triangles as possible in the diagram. You can direct her first to count all the small individual triangles, then to try to find larger triangles made up of 4 small triangles (there are 6 of them).

There are also different shapes that are made up of the small triangles. There are many different diamond shapes made up of two triangles each. There are also 2 large diamonds, each made up of 8 small triangles. There are many different trapezoids, and there are 3 different hexagons. Have the child find as many different shapes as she can. If you have several copies of the diagram, you can direct the child to color in the various shapes as they are found. You can return to this many times and each time find something new.

Find the Squares

SKILL AREA: Spatial Visualization

GRADE LEVEL: Intermediate, junior high

STRATEGIES: Success for each child
            Many right answers
            Many approaches
            Independent work

MATH CONCEPTS: Small squares can fit together to make larger squares.

The smallest square is 1 square unit; the next larger square

66

is 4 square units; the next larger square is 9 square units. The numbers of square units needed to make larger squares are known as *square numbers:* 1, 4, 9, 16, 25, etc.

A table can be used to organize information about squares.

Patterns can be found in the table to give information about larger squares.

MATERIALS: Squared paper or diagrams of squares as shown, pencils.

DIRECTIONS: The first part of this activity involves having the child look at the diagram and find as many hidden squares as she can. At first, she will do this randomly, then she will start to develop a system for finding all the squares in a diagram. For example, in this square there are really 5 squares, the 4 small squares and the large square made up of the smaller ones.

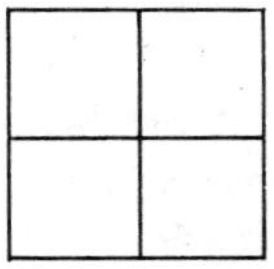

FIGURE 5-8.

Can you find the 14 hidden squares?

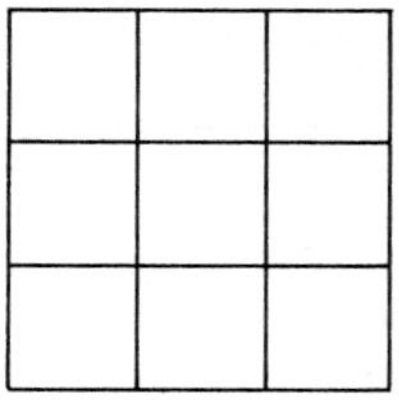

FIGURE 5-9.

Can you find the 30 hidden squares?

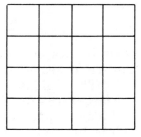

FIGURE 5-10.

Can you find the 55 hidden squares?

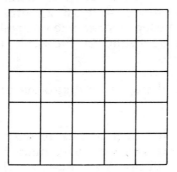

FIGURE 5-11.

After finding the hidden squares in these diagrams, an older child may want to organize this information in a table. She can then look for patterns in the table and find a rule which will tell how many squares can be found in the next larger square. For example, if you have a square that is 7 by 7 (made up of 49 small squares), the rule will tell how many hidden squares there are; it would not be necessary to draw the pictures and count all the squares.

This table organizes the information.

Table 5-1. Hidden Squares Pattern

Dimensions of the Square	*Number of Hidden Squares*
1 x 1	1
2 x 2	5
3 x 3	14
4 x 4	30
5 x 5	55
6 x 6	————
7 x 7	————
8 x 8	————
9 x 9	————
10 x 10	————

Different patterns can be found which connect these numbers and tell what number comes next. These patterns can be explained in various ways. The child is to find these patterns and fill in the blanks for the larger squares. For example, here is one way of finding all the hidden squares for a 6 by 6 square: first multiply 6 by 6, which is 36, then add 36 to the previous total, 55, for the 5-by-5 square. Thus, there are 91 hidden squares in a 6-by-6 square.

5-Square Geometry

Making Pentominoes

SKILL AREA: Spatial Visualization

GRADE LEVEL: Primary, intermediate, junior high

STRATEGIES: Success for each child
Many right answers
Using manipulatives
Cooperative work

MATH CONCEPTS: *Congruent* figures are the same size and shape; they fit on top of each other; and are congruent shapes.

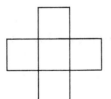

Different shapes can have the same *area*. Both and have an area of 5 square units.

MATERIALS: Each child needs 5 cardboard squares, 2 in. by 2 in. (or 5 cm by 5 cm); some construction paper, 12 in. by 18 in. (or 30 cm by 45 cm); scissors; a pencil.

DIRECTIONS: Shapes that are made out of 5 squares are called pentominoes. The whole side of each square must touch a whole side of another square. This kind of arrangement is correct.

FIGURE 5-12.

These arrangements are incorrect.

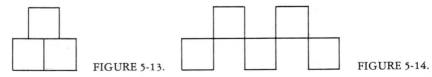

FIGURE 5-13.   FIGURE 5-14.

Ask children to arrange their 5 squares into a pentomino shape, then trace around each square on construction paper, to show both the lines of the individual squares and the pentomino shape. Have the children draw as many different pentominoes as they can, then cut them out. Here are the 12 different pentominoes that can be formed.

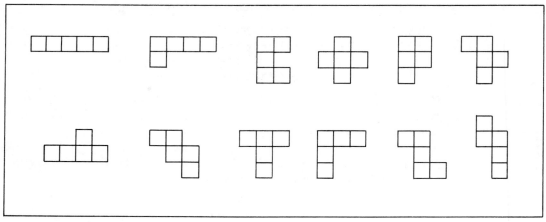

FIGURE 5-15.

In this activity there is an opportunity for each child to be successful. Everyone will easily find some of the 12 possible shapes. In a classroom, make finding all the shapes a group rather than an individual task. Children can return to this activity several times in order to find all the shapes.

Folding Boxes

SKILL AREA: Spatial Visualization

GRADE LEVEL: Primary, intermediate, junior high

STRATEGIES: Guessing and testing
Success for each child
Using manipulatives
Many right answers

MATH CONCEPTS: Most boxes have 6 faces.
Three-dimensional structures can be constructed from two-dimensional shapes; for example, cubes can be constructed from squares.

MATERIALS: A set of 12 pentominoes, a few boxes of different sizes.

DIRECTIONS: Show children various closed boxes. Many think a box has only 4 faces. Have them count the faces on the boxes. Ask them how many faces the boxes would have if they had no tops on them.

In this activity the children look at their pentominoes and decide which can be folded to make open boxes. They can check their guesses by actually folding the pentominoes. Only 8 of the 12 shapes can be folded into boxes.

Guessing and testing is the method the children will use. One approach to doing this activity is to discuss 3 of the shapes at a time. For example, ask, "Do you think this shape can be folded to make an open box? Write down 'yes' or 'no'." Do the same for 2 more of the shapes. Then the children check these guesses by actually folding the pentominoes to see if they do form an open box. Continue taking groups of 3 pentominoes until they have tried all 12 shapes. By the last group there will be increased confidence among the children. They really won't be "guessing" anymore; they will, in fact, know the answers.

Milk Carton Cutting

SKILL AREA: Spatial Visualization

GRADE LEVEL: Primary, intermediate, junior high

STRATEGIES: Guessing and testing
Success for each child
Using manipulatives
Many right answers
Cooperative work

MATH CONCEPT: Three-dimensional figures can be cut apart to show their two-dimensional components.

MATERIALS: A large supply of milk cartons with the tops cut off (quart or school size is preferable, cut so that the faces are approximately the same size), scissors

DIRECTIONS: The object of this activity is to cut the milk carton so that it flattens out to make one of the pentomino shapes. You can only make the 8 shapes that can be folded into boxes. Let each child try several cartons, so that she can make different shapes. Provide at least 3 or 4 milk cartons for each child. This will enable children to guess and test, to make mistakes, and to try again. Making this into a group task alleviates pressure on an individual child to cut all 8 patterns.

The Last Piece
(*a strategy game using pentominoes*)

SKILL AREA: Spatial Visualization

GRADE LEVEL: Primary, intermediate, junior high

STRATEGIES: Using manipulatives
Game playing

MATH CONCEPTS: *Congruent* figures are the same size and shape. Shapes can be rotated or turned over to fit into a certain space.

MATERIALS: A set of 12 pentominoes made out of 2-in. (or 5cm) squares, a checkerboard with 2-in. (or 5cm) squares

DIRECTIONS: This is a two-player strategy game with rules that are easy to learn. Spread out the pentominoes on a table so that the players can easily see them. On her turn, each player chooses one of the pentomines and places it on the checkerboard. The last player that can place a shape on the board is the winner. At first the game is played randomly, but in a very short time children learn to visualize a move before choosing their pentomino. They also start to plan their moves in advance, attempting to block their opponent.

Match the Faces

SKILL AREA: Spatial Visualization

GRADE LEVEL: Primary, intermediate

STRATEGIES: Using manipulatives
       Success for each child
       Guessing and testing
       Estimating
       Independent work

MATH CONCEPTS: Most boxes have 6 faces, which are either squares or rectangles.

MATERIALS: Several (about 6) small boxes of various shapes and sizes, wrapped in wrapping paper (matchboxes, check boxes, jewelry boxes, and spice boxes would work well); construction paper; pencils

DIRECTIONS: In the first part of this activity, each child is to trace the faces of one box onto a piece of construction paper. Marking each face on the box as she traces it will help her to keep track of the faces. Each child's paper will look something like this for each box.

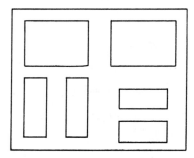

FIGURE 5-16.

The next part of the activity involves matching the correct box with the paper that has the faces drawn on it. One way of approaching this task is to place the several boxes in a row on the table. Have all the drawings of the faces in a pile in a random order. The child is to look at the faces on the top paper, then choose a box that might fit those outlines. Then, she must actually place the box on the paper to make sure that all the faces match the outlines. If they don't all match, she must try another box. Repeat this process until all the boxes are matched with the traced faces.

A child and parent can work at this activity in pairs. They can take turns choosing a box and making the outlines without the other person watching. Then the other person must decide which box fits the outline.

Building with Cubes

Exploring and Counting

SKILL AREA: Spatial Visualization

GRADE LEVEL: Primary, intermediate

STRATEGIES: Using manipulatives
Many right answers
Cooperative work

MATH CONCEPTS: A *cube* is a three-dimensional object with 6 congruent square faces.
Volume is measured in cubic units.
Different-looking structures can have the same volume.

MATERIALS: A large supply of small wooden cubes or sugar cubes

DIRECTIONS: Give each child 24 cubes with which to build a structure. If one child is doing this activity, let her build several structures that each use 24 cubes. Then examine the different structures that the children have built. Some structures may have more than one layer. Some structures may be rectangular. All the structures have the same volume, since they were built with the same number of cubes.

Later, direct the children to build only certain kinds of structures. How many different rectangular shapes can they make? Can they make rectangular shapes that have more than one layer?

As an extension of the activity, have the children look at a structure and figure out how many cubes were used to build it. Children can work in pairs for this task. One child builds her own structure using any number of cubes. The other child figures out how many cubes were used. They take turns being the builder or the counter. The children will not simply be able to count the cubes they can see, as some cubes may be hidden.

Describing Structures

SKILL AREA: Spatial Visualization

GRADE LEVEL: Intermediate, junior high

STRATEGIES: Using manipulatives
Success for each child
Many approaches
Many right answers
Cooperative work

MATH CONCEPTS: Three-dimensional rectangular structures can be described in terms of their length, width, and height.
Different-looking structures can have the same volume.
Structures that have the same volume may have different surface areas.

MATERIALS: A large supply of cubes

DIRECTIONS: Two children can take turns in this activity. One child builds a rectangular structure, then describes it to her partner, who has not seen it. The second child must build the structure according to the description. Afterwards they switch roles. This shape could be described as being 2 cubes wide, 4 cubes long, and 3 cubes high (or have 3 layers).

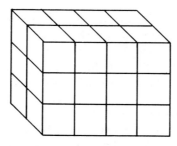

FIGURE 5-17.

Alternately, prepare some cards that describe the structures. Then have the child choose a card and build that specific structure.

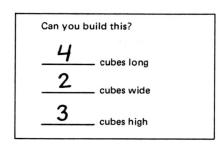

FIGURE 5-18.

Marking Faces

DIRECTIONS: In this activity, the concept of surface area is informally introduced. Children are to figure out the surface area of several structures that have been built with the same number of cubes.

Direct each child to build a structure using 24 cubes. A single child may build several different structures. Have them mark an X on each face of each cube of their structure that is showing on the outside surface. Then have them count the number of Xs. Although the structures were all built with 24 cubes, the number of faces on them will vary; therefore, the surface areas are not the same.

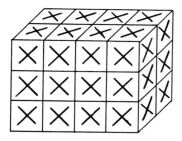

FIGURE 5-19.

Challenge children to make a structure that has more Xs or fewer Xs than the one they have already made.

Making a Mobile

SKILL AREA: Spatial Visualization

GRADE LEVEL: Primary, intermediate

STRATEGIES: Success for each child
                        Using manipulatives

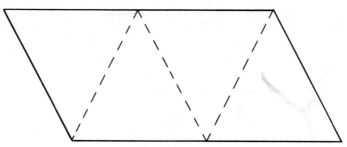

FIGURE 5-20. Triangular pyramid.

FIGURE 5-21. Rectangular prism.

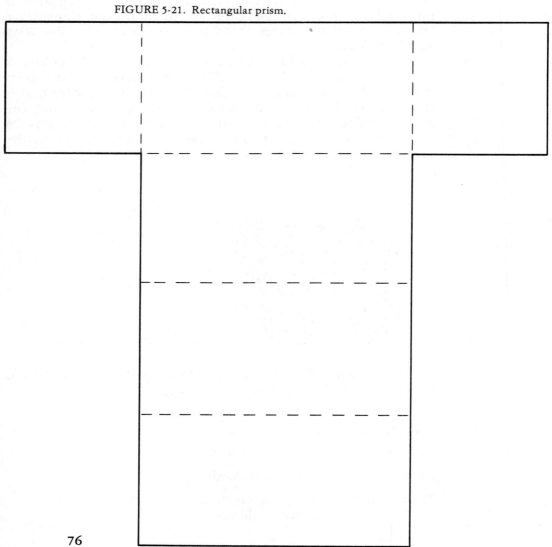

MATH CONCEPTS: *Cube, rectangular prism, square pyramid, triangular pyramid,* and *triangular prism* are common solid figures.

The *faces* of these solids are squares, rectangles, or triangles. These solids have 4, 5, or 6 faces.

The line formed by connecting two faces is called an *edge.*

The corner points are each called a *vertex.*

MATERIALS: Patterns of the solid figures, scissors, tape, needle and thread, a wire hanger

DIRECTIONS: Children can easily cut out and fold these patterns to make the different solids. Have them trace the patterns onto heavier paper, cut on the solid lines around the outside, and fold on the dotted lines. Before taping, they should draw a needle and thread through one of the faces of each figure. The threads are knotted around the wire hanger to make the mobile.

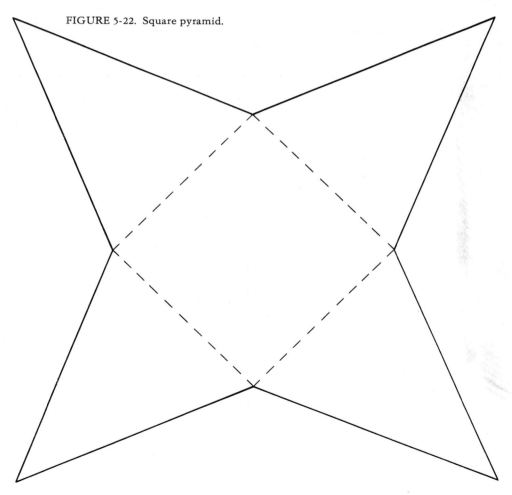

FIGURE 5-22. Square pyramid.

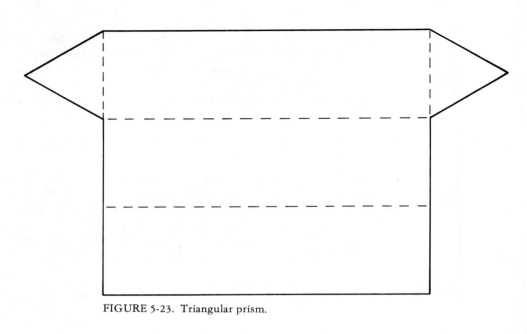

FIGURE 5-23. Triangular prism.

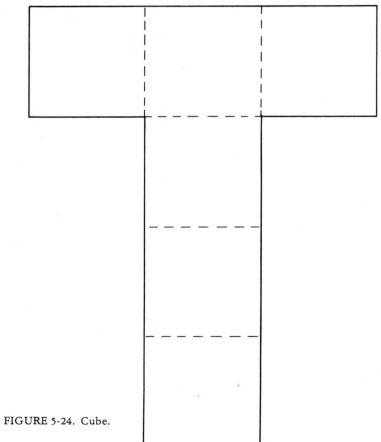

FIGURE 5-24. Cube.

The following activities provide practice in locating points on a grid. Young children can easily learn this skill, which is used in many areas of mathematics.

SKILL AREA: Spatial Visualization

GRADE LEVEL: Primary, intermediate

STRATEGIES: Success for each child
            Game playing
            Many right answers
            Independent work

MATH CONCEPTS: An *ordered number pair,* such as (2,1), names a specific point on a grid.

The first number directs you to move horizontally, and the second number directs you to move vertically on the grid in order to locate the point. For example, notice the location of the points (2,1) and (1,2).

The order of the numbers is important; (1,2) is not the same point as (2,1).

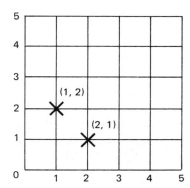

FIGURE 5-25.

Three-in-a-Row

MATERIALS: One 5-by-5 gameboard, easily drawn on paper, used by both players; a different-colored crayon for each player; 10 cards made up as follows:

FIGURE 5-26.

OVER 1	OVER 2	OVER 3	OVER 4	OVER 5
UP 1	UP 2	UP 3	UP 4	UP 5

DIRECTIONS: Two children take turns in this game. Keep the "over" and "up" cards face down in separate piles. On each turn, a player draws one card from each pile. Then she locates the point determined by the cards and marks an X on that point. Finally, she returns the cards to their piles and mixes them in with the other cards. If a player is to locate a point that has already been marked, she loses that turn. The winner is the player who marks 3 points in a row either horizontally, vertically, or diagonally.

On the sample gameboard which follows, the point has been marked for the cards labeled "OVER 4" and "UP 2." Start at the zero, move over to the number 4, then move up to the number 2.

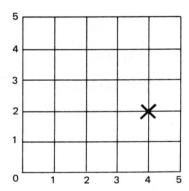

FIGURE 5-27.

Add-a-Coin

MATERIALS: A gameboard made up of a 5-by-5 grid with the values of coins written on the intersections, the 10 cards used in Three-in-a-Row, a supply of coins

DIRECTIONS: This game for two or three players gives children an opportunity to practice recognizing coins and their values and adding up the values of the coins. Players take turns drawing an "over" and "up" card, locating the point determined by the cards, and collecting the coin that has the same value as the amount of money on the point. After each card is used, it is returned to the pile and mixed in with the other cards. The first player to collect a certain amount is the winner.

The game can easily be changed for different grade levels. You could use only pennies and nickels and have the winning total be 25¢. Then, you could include dimes and change the total to 50¢. Later, quarters and half-dollars could be added and the total changed to $1.00.

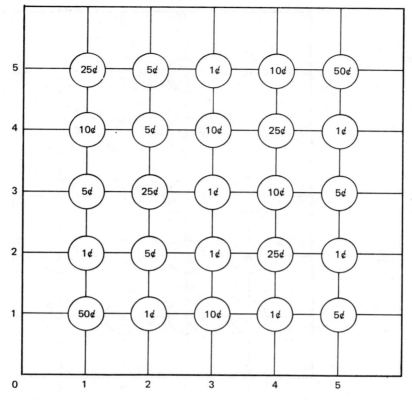

FIGURE 5-28.

Secret Codes

MATERIALS: Pencils, a 5-by-5 grid with 25 of the 26 letters of the alphabet printed on the intersection points (any letter may be omitted; words with that letter cannot be used)

DIRECTIONS: The secret code is made up by using the coordinate points as the code for the letters on the grid. For example, on the grid shown, the code for the letter W is (2,3), the code for the letter G is (4,1), and the code for the letter F is (3,2). Below the grid, write the number pairs of the code, leaving a blank above each one.

To read the message, the child must find the letter that goes with each number pair and write it on the blank. Children soon will be able to make up their own messages and enjoy exchanging them with their friends.

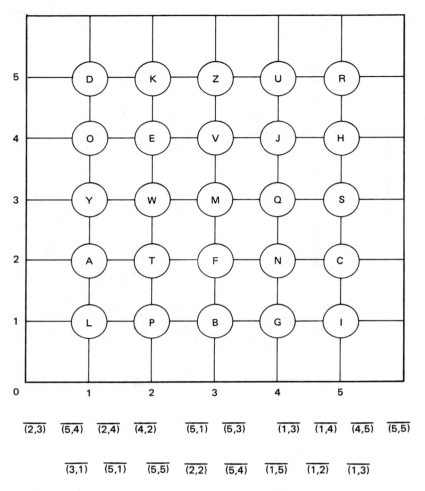

FIGURE 5-29.

$\overline{(2,3)}$ $\overline{(5,4)}$ $\overline{(2,4)}$ $\overline{(4,2)}$ $\overline{(5,1)}$ $\overline{(5,3)}$ $\overline{(1,3)}$ $\overline{(1,4)}$ $\overline{(4,5)}$ $\overline{(5,5)}$

$\overline{(3,1)}$ $\overline{(5,1)}$ $\overline{(5,5)}$ $\overline{(2,2)}$ $\overline{(5,4)}$ $\overline{(1,5)}$ $\overline{(1,2)}$ $\overline{(1,3)}$

Make a Picture

MATERIALS: A 10-by-10 grid with number pairs which connect to form a picture, pencils

DIRECTIONS: In this activity, the child forms a picture on the grid by connecting the points that are named by the number pairs. It is important to connect the points in the order in which you find them. Direct the children to find the first point and mark it with an X; then find the second point, mark that with an X, and connect the two points. Continue finding the points and connecting them to the previous point until the picture is formed. Children may be encouraged to make up their own pictures and list the points they need to make it. Here are a few samples. The first one is started for you.

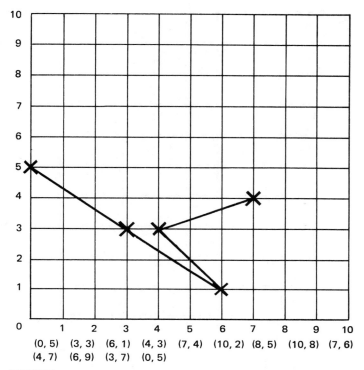

(0, 5) (3, 3) (6, 1) (4, 3) (7, 4) (10, 2) (8, 5) (10, 8) (7, 6)
(4, 7) (6, 9) (3, 7) (0, 5)

FIGURE 5-30.

FIGURE 5-31.

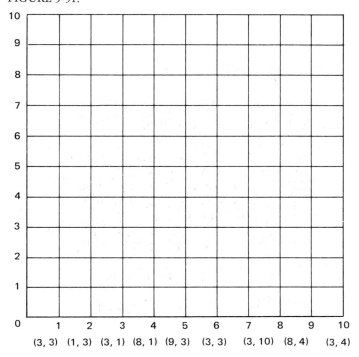

(3, 3) (1, 3) (3, 1) (8, 1) (9, 3) (3, 3)   (3, 10) (8, 4)   (3, 4)

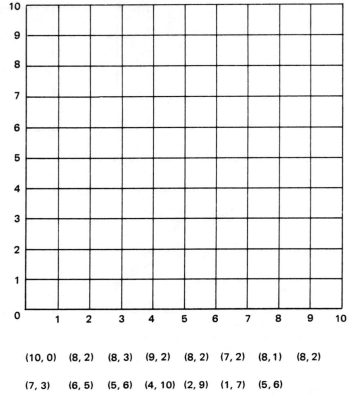

(10, 0) (8, 2) (8, 3) (9, 2) (8, 2) (7, 2) (8, 1) (8, 2)

(7, 3) (6, 5) (5, 6) (4, 10) (2, 9) (1, 7) (5, 6)

FIGURE 5-32.

Making Graphs

The following three graphing activities are included to provide an introduction to graphing for children at different levels. The Birthday Graph, on which cards are used to represent each piece of information, is the most concrete of the graphs. The Ice Cream Graph is more abstract, with each square representing an ice cream cone. The Temperature Graph, the most abstract, is a line graph. These activities fit well with the graphing activities presented in the Grid Games. They are also useful as preparation for the graphing used in the science activities in chapters seven and eight.

SKILL AREA: Spatial Visualization

84   GRADE LEVEL: Primary, intermediate

STRATEGIES: Success for each child
Many right answers
Using manipulatives
Independent work

MATH CONCEPTS: Making a graph is a way to organize information.
Graphs communicate information by showing relationships.

Birthday Graph

MATERIALS: A sheet of plain paper, about 48 in. by 36 in. (120cm by 90cm); index cards or cards cut from colored paper, about 2 in. by 3 in. (5cm by 8cm); a marking pen; tape

DIRECTIONS: Mark 12 columns on the large paper, each about 4 in. (10cm) wide. Print the months of the year at the bottom of the columns. The child is to fill the small cards with the names and birthdates of as many people as possible. She then tapes each card in the appropriate column on the graph. In this simple graph, no numbers are necessary on the vertical axis. By looking at the results a young child can answer many questions such as:

In which month do we have the most birthdays?

Are there any months in which we don't have any birthdays?

Counting skills are reinforced by questions like:

How many people have birthdays in June? in September?

Are there any months in which two people have birthdays?

Simple computation problems can be included with questions like:

How many birthdays are there altogether in January and March?

How many more birthdays are there in June than August?

The results on the graph will determine the exact questions that you make up.

In a classroom, each child writes his or her own birthdate on a small card, then tapes it in the correct place on the class graph. Discussion of the information should follow. Children can also make up their own questions and statements about the information displayed.

		JOAN						JANE			
		CAROL						ALLEN			
		DAVID						TED			
		BOB						ED		BILL	
MARY		ELAINE				CAROL					
ANN		JUDY		LUCY		SUE		DIANE			
Jan.	Feb.	March	April	May	June	July	Aug.	Sept.	Oct.	Nov.	Dec.

FIGURE 5-33. Birthday graph.

Ice Cream Graph

MATERIALS: A large sheet of graph paper; 8½-in.-by-11-in. (20cm-by-30cm) graph paper (1-in. or 2.5cm squares) for each child; crayons; pictures of ice cream cones in four different flavors

WHICH IS YOUR FAVORITE FLAVOR ?

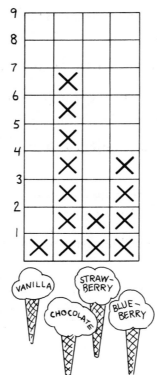

86

FIGURE 5-34. Ice cream graph.

DIRECTIONS: In this activity children can fill in their individual graphs as a large class graph is being completed. Prepare this graph by making four columns with an ice cream flavor pictured at the bottom of each one. Write the numerals on the vertical axis. Individual student papers should look similar to the large graph. The graph will be more interesting if you limit the choices that the children can make. Without doing this, your results might be 20 different favorite flavors, each chosen by one or two children. In a classroom, each child marks with an X or colors in one square in the column above the picture of her favorite flavor. At home, a child can do this as a survey, asking people which flavor is their favorite and coloring the appropriate squares on the graph. Whether at home or in school, look at the results and encourage children to ask questions as well as to answer them.

Temperature Graph

SKILL AREA: Spatial Visualization

GRADE LEVEL: Intermediate, junior high

STRATEGIES: Success for each child
Independent work

MATH CONCEPTS: Making a graph is a way to organize information.

Graphs communicate information by showing relationships.

The *mean* is the average temperature. It is found by adding up all the temperatures, then dividing the total by the number of readings that were taken. In the sample shown, the mean temperature is 47 degrees ($42 + 48 + 46 + 50 + 46 + 46 + 51 = 329$; $329 \div 7 = 47$).

The *median* is the middle temperature when the readings are ranked in order. The median in the sample is 46 degrees.

The *mode* is the temperature that was graphed the most times. On three days the temperature was 46, so the mode is 46 degrees.

The *range* is the difference between the lowest and highest reading. The temperatures in the sample go from 42 to 51 degrees; the range is therefore 9 degrees.

MATERIALS: A large sheet of graph paper, pencils, rulers

DIRECTIONS: Prepare a large graph by writing the dates for one week along the horizontal axis and the temperature in degrees on the vertical axis. Each day at the same hour, have a child take the temperature at a specific location outdoors. To mark the graph she must first move horizontally to the correct date, then vertically to the correct temperature, and mark the intersection point with an X. She should draw a line connecting her mark to the previous recording. At the end of the week, children can discuss the results shown on the graph. It would be useful to make two graphs, one using Fahrenheit and the other using Centigrade. Children can find the *mean, median, mode,* and *range* for the information on each graph.

FIGURE 5-35. Temperature graph.

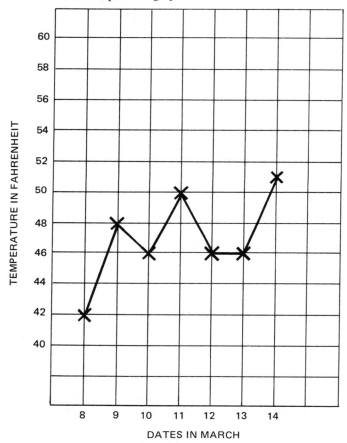

SCIENCE ACTIVITIES

Practice in spatial tasks helps prepare children for later work in science, as well as mathematics. This is because spatial visualization is essential in both the biological and physical sciences. For example, biologists visualize cells and organelles, geologists visualize rock formations and read contour maps, chemists visualize molecules and molecular interactions, and astronomers visualize the configurations of bodies in space.

Scientists routinely construct and interpret two-dimensional representations—graphs, charts, maps, drawings—of three-dimensional objects. They also construct three-dimensional models of objects that are too small or too large to interact with directly, such as molecules and planets. These models help scientists clarify, communicate, and test their ideas regarding shapes and structures.

In addition to visualizing objects, scientists visualize problems. Research has shown that in solving physics problems, experts begin by visualizing the physical situation, whereas novices try immediately to produce and use equations.[1]

The skills developed in the science activities on spatial visualization are:

1. recognizing shapes and their relationships to each other;

2. imagining what objects would look like rotated or sliced through the middle;

3. identifying planes of symmetry;

4. making and interpreting models.

By manipulating objects and by taking objects apart and putting them together, boys tend, more than girls, to develop these skills during play.

The first activity, Noticing Shapes, deals with recognizing shapes and seeing how they are put together in an object. These skills are critical to spatial visualization. If we cannot identify shapes and see their relationships to each other, we cannot reproduce structures in the imagination. Young children's stereotypical drawings of trees and houses suggest that they miss many important aspects of the objects they observe. Noticing Shapes requires children to identify shapes and relationships in trees, leaves, and houses, and to remember and use what they observe.

Making Shadows gives children experience in visualizing rotations and slices through objects. By rotating an object, they can produce many different shadows. Some of the shadows they make are silhouettes of cross-sections through simple objects. The sequence of activities also gives them the opportunity to make more complex shadows, such as animal shadows using their own hands.

---

[1]Jill H. Larkin, "Processing Information for Effective Problem Solving" (unpublished research paper, presented at the winter meeting of the American Association of Physics Teachers in Chicago, Illinois, February 1977), p. 5.

A plane of symmetry is a special kind of cross-section. In Exploring Symmetry, children learn about symmetry by identifying patterns that could be made by holding a mirror next to a simple shape, and by using mirrors to make their own symmetrical designs.

In Carving Canyons and Making Molecules, children construct models of rock formations and simple chemical compounds. These models are similar to those a scientist would construct. Boys, who generally make model airplanes, cars, and buildings, are much more likely than girls to develop proficiency in model construction and interpretation in their normal activities.

Noticing Shapes

In these activities, a child learns the names of shapes, such as oval and triangle, and finds the shapes in the things she observes during a walk.

SKILL AREA: Spatial Visualization

GRADE LEVEL: Primary

STRATEGIES: Using manipulatives
Success for each child
Many right answers
Independent work

SCIENCE CONCEPT: Objects with similar functions come in many different shapes.

MATERIALS: Paper, crayons, shapes cut from construction paper of various colors, magazines, scissors, paste, shallow containers to hold pictures cut from magazines

Shapes on Houses

DIRECTIONS: Show the child the following shapes cut from construction paper: oval, triangle, rectangle, arch, square. Have her identify the shapes if she can; otherwise, tell her their names. Ask her to find these shapes in magazine pictures, cut them out, and place them in appropriately labeled containers.

Ask the child to find ovals, triangles, rectangles, arches, and squares on houses. (This can be done in a classroom using pictures of houses.) What other shapes can she find? Afterwards, have her draw a house, using the shapes she has learned.

DIRECTIONS: Use shapes cut from construction paper to introduce "circle," "umbrella," and "cone."

Can the child find trees with crowns shaped like circles, umbrellas, and cones? What kinds of trees does she see? What other shapes can she find? Have her draw a cluster of trees of various shapes.

Shapes of Leaves

DIRECTIONS: Use shapes cut from construction paper to introduce "heart," "fan," and "needle."

Have the child collect leaves. Can she find leaves with these shapes: heart, fan, needle, oval, circle, fingers? What shapes can she find? How would she describe the leaves she sees? Have her paste her leaves on a large sheet of construction paper and write the names of the shapes underneath.

Making Shadows

If these activities are done outdoors in sunlight, children can make shadows on the ground. If they are done indoors, a flashlight or projector can be used as a light source, and children can make shadows on a screen or light-colored wall.

What Is a Shadow?

SKILL AREA: Spatial Visualization

GRADE LEVEL: Primary

STRATEGIES: Using manipulatives
Success for each child
Many right answers
Independent work

SCIENCE CONCEPT: To make a *shadow,* you need a light and an object that blocks the light.

MATERIALS: Light sources (flashlight, electric light, mirror to reflect light), miscellaneous objects (paper cups, toy balls, buttons, pencils, straws, tubes, rectangular boxes, coins, ice cream cones, etc.)

DIRECTIONS: Ask children how shadows are made. Is it possible to make a shadow without a light? What is necessary besides a light? What kinds of light sources will make shadows? Have the children generate a list: sun, moon, candle, electric light, flash-light, mirror reflecting light, etc. Let children experiment, making shadows with various objects and light sources.

Cone Shadows

SKILL AREA: Spatial Visualization

GRADE LEVEL: Primary, intermediate

STRATEGIES: Using manipulatives
                Success for each child
                Guessing and testing
                Independent work

SCIENCE CONCEPT: A single object can have many different shadows.

MATERIALS: Light source, scissors, tape or stapler, copies of cone pattern

DIRECTIONS: Give each child a copy of the following pattern:

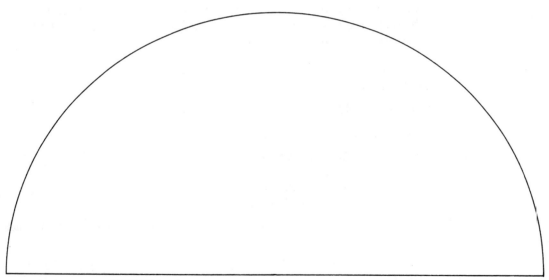

FIGURE 5-36.

Have them cut out the figure, roll it into a cone, and secure it with tape or staples:

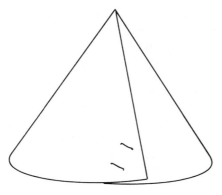

FIGURE 5-37.

Tell them the top of the cone should be completely closed, not open like a volcano.

Draw the following shapes on a chalkboard or point to them in the book. Ask children to guess which ones could be shadows made by the cone.

FIGURE 5-38

A. Circle

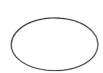

B. Ellipse

C. Cone

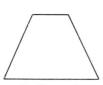

D. Trapezoid

E. Rectangle

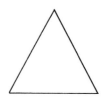

F. Triangle

H. Tear

Let them experiment to test their guesses. Afterward, they can vote on whether each of the shapes can be made; volunteers can demonstrate how to make them.

The trapezoid and rectangle cannot be made unless the shape of the cone is destroyed by folding. All of the other shapes can be made. Make sure that the children see that there are two ways to make the circle and several ways to make the ellipse: the circle can be made by pointing the cone either directly toward or directly away from the light source; the ellipse can be made the same way as the circle, but compressing the cone slightly, or by holding the cone at a slight angle.

Similar Shadows

SKILL AREA: Spatial Visualization

GRADE LEVEL: Primary, intermediate

STRATEGIES: Using manipulatives
Success for each child
Many right answers
Guessing and testing
Independent work

SCIENCE CONCEPT: Unrelated objects can make similar shadows.

MATERIALS: Light source, miscellaneous objects

DIRECTIONS: Show children a collection of objects, and ask them to find pairs of objects that can make similar shadows. For example, a cone and a coin can both make circular shadows. Let them test their guesses. Older children can list successful pairs on the chalkboard, each pair with a drawing of the similar shadow below.

Animal Shadows

SKILL AREA: Spatial Visualization

GRADE LEVEL: Primary, intermediate

STRATEGIES: Using manipulatives
Success for each child
Many right answers
Independent work
Cooperative work

SCIENCE CONCEPT: To make a *shadow,* you need a light and an object that blocks the light.

MATERIALS: Light source, paper, pencils

DIRECTIONS: Challenge children to use their hands to make shadows that look like the following animals: a rabbit, a duck, a turkey, an alligator, a dog. What other animals can they make? The shadows can be made on white paper and traced by a helper. This makes a fun homework assignment.

Changing Shadows

SKILL AREA: Spatial Visualization

GRADE LEVEL: Intermediate

STRATEGIES: Using manipulatives
         Success for each child
         Many right answers
         Guessing and testing
         Independent work

SCIENCE CONCEPT: The size, shape, and sharpness of a shadow can be changed by moving the light, the object casting the shadow, or the surface on which the shadow falls.

MATERIALS: Light source, miscellaneous objects

DIRECTIONS: Ask children to find 3 ways to change each of the following aspects of a shadow: size, sharpness, shape. After they have had time to experiment, let volunteers demonstrate the following propositions:

> A shadow can be made smaller and sharper (1) by moving the object casting the shadow closer to the surface on which the shadow falls, (2) by moving the surface on which the shadow falls closer to the object casting the shadow, or (3) by moving the light farther away from the object casting the shadow.

> The shape of a shadow can be changed (1) by rotating the object casting the shadow, (2) by tilting the surface on which the shadow falls, or (3) by moving the light source.

Predicting Shadows

SKILL AREA: Spatial Visualization

GRADE LEVEL: Intermediate

STRATEGIES: Using manipulatives
Guessing and testing
Independent work

SCIENCE CONCEPT: A single object can have many different shadows.

MATERIALS: Light source, miscellaneous objects

DIRECTIONS: Demonstrate how to rotate an object through 1/8, 1/4, 3/8, 1/2, and 3/4 turn. Let children predict the shadows that would be produced by rotation of the object you are holding. Volunteers can draw the predicted shadows on the chalkboard and test the predictions.

Give each child a small object. Tell children to draw a shadow they think they could make with their object. Ask them to think about what would happen to the shadow as they rotated the object and to draw the different images that would be formed. Let them test their predictions.

Exploring Symmetry

Matching Patterns

SKILL AREA: Spatial Visualization

GRADE LEVEL: Primary, intermediate

STRATEGIES: Using manipulatives
Success for each child
Guessing and testing
Independent work

SCIENCE CONCEPTS: A *mirror* is a smooth, shiny surface that forms *images*.

The image in a mirror is *reversed*.

MATERIALS: Pocket mirrors, copies of mirror patterns shown here

DIRECTIONS: Give each child a mirror and a set of mirror patterns (at home, your child can use the figures in the book):

FIGURE 5-39.

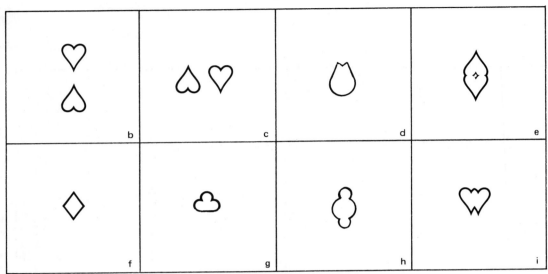

FIGURE 5-40.

FIGURE 5-41.

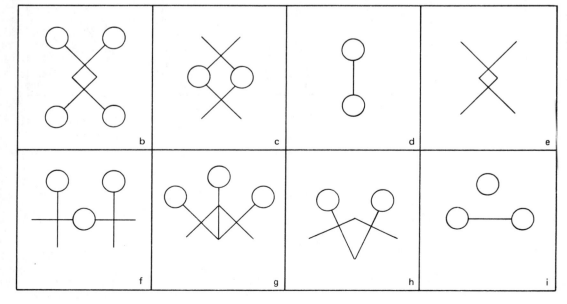

Draw the following figure on a chalkboard or point to it in the book:

FIGURE 5-42.

Ask children if, by holding the mirror next to the figure, they can make this pattern:

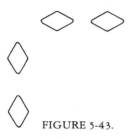

FIGURE 5-43.

When everyone has tried it, let the children who were successful show the others how to do it.

Challenge children to use their mirrors to make the other diamond patterns. Tell them that some of the patterns cannot be made. Emphasize that it's okay to try a lot of guesses. Check the children as they work; some might think they have made the patterns when they actually have not. After they have had time to work independently, let the children discuss the patterns and show each other how to make them. Diamond patterns (f) and (i) cannot be made.

Repeat the above procedure with the heart and the lollipops. The following patterns cannot be made: heart (c) and (h), lollipops (g).

At the end of the activity, ask children how a mirror changes the appearance of an object. Introduce the word *reversal,* but encourage the children to discuss the concept using their own words.

Mirror Art

SKILL AREA: Spatial Visualization

GRADE LEVEL: Primary, intermediate

STRATEGIES: Using manipulatives
Success for each child
Many right answers
Independent work

SCIENCE CONCEPTS: A *mirror* is a smooth, shiny surface that forms *images*.
The image in a mirror is *reversed*.

MATERIALS: Pocket mirrors, pencils, crayons, white paper

DIRECTIONS: Children enjoy making and copying their own designs using mirrors and simple shapes, such as the diamonds, heart, and lollipops in Matching Patterns. This is a nonthreatening activity that helps develop spatial skills.

Make one or two samples to show the children what they are expected to do:

> Fold a piece of white paper into eighths and open it again. In the upper left-hand box, draw any simple shape. Experiment making designs by holding a mirror next to this shape. In each of the remaining boxes, draw a pattern made with the mirror and the original shape. Color your designs.

If children have not done Matching Patterns, explain how you made your designs. You can either provide children with starting shapes or let them draw their own.

Carving Canyons

SKILL AREA: Spatial Visualization

GRADE LEVEL: Intermediate, junior high

STRATEGIES: Using manipulatives
Cooperative work

SCIENCE CONCEPTS: The Earth's crust contains many different kinds of *rock formations* (see Figure 5-45).

A rock formation can make a *trace* across a stream valley:

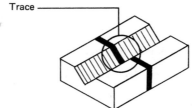

FIGURE 5-44.

FIGURE 5-45. Basalt dike cutting through granite.

The trace across a stream valley tells a geologist the direction (north, south, east, west, etc.) and amount of tilting of a rock formation.

MATERIALS: Modeling clay (2 contrasting colors), plastic knives, rolling pins (optional), pencils, crayons, paper, clay models of rock formations and canyon (instructions follow)

DIRECTIONS: Before starting this activity, you will need to make three clay models of rock formations and one of a canyon:

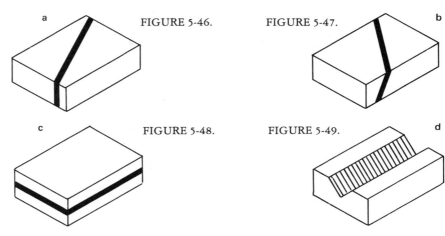

FIGURE 5-46.

FIGURE 5-47.

FIGURE 5-48.

FIGURE 5-49.

Make the models approximately 2½ in. by 2½ in. by ¾ in. (6cm by 6cm by 2cm). You can construct models (a) and (b) most easily by making the basic form in one color of clay, cutting it in two, inserting a contrasting sheet of clay, and trimming the edges. To make the tilt in (b), hold the knife at an angle with respect to the surface of the model. Make model (c) by rolling out three sheets of clay, stacking them, and trimming the edges. For model (d), use only one color of clay; carve the canyon with a plastic knife.

Tell children that the Earth's crust contains many different kinds of rock formations. Let them look at pictures of rock formations and canyons. Then show them models (a), (b), and (c), and explain that each color of clay represents a different kind of rock. Point out that in model (a), the dark strip is vertical, whereas in model (b), the dark strip is tilted.

Show the children model (d), and tell them that the groove represents a stream valley. Ask them to imagine carving similar canyons through the other three models. What would the darker rock look like where it crossed the canyon? Pass the models around for children to examine.

Working in groups of three, have children construct their own set of models (a, b, and c) and carve the canyons. When they have finished, ask them what they think geologists learn from the rock patterns on canyon walls. They should mention that (1) different kinds of rock are present, (2) the formations run in different directions (for example, north to south or east to west), and (3) the formations show different amounts of tilting.

Ask children to draw diagrams of their models as seen from straight overhead. They should look at their models carefully as they work. After they have done this, draw the *geologic maps* on the chalkboard. (See Figure 5-50, right column.)

Have the children guess which map matches which model. Do their drawings look like the maps?

Ask the children how they think a rock formation becomes tilted. (Some are formed that way, whereas others become tilted due to earthquakes or gradual movement of the land.)

Making Molecules

SKILL AREA: Spatial Visualization

GRADE LEVEL: Intermediate, junior high

STRATEGIES: Using manipulatives
Independent work

SCIENCE CONCEPTS: All matter (solids, liquids, gases) is composed of tiny particles called *atoms.*

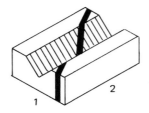

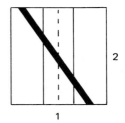

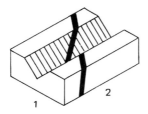

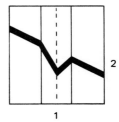

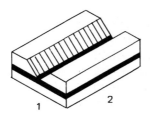

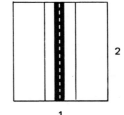

FIGURE 5-50.

Atoms are held together by *bonds*.

A *molecule* consists of two or more atoms held together by bonds.

Simple Molecules

MATERIALS: Wooden toothpicks, modeling clay (3 different colors), clay model of human figure, clay model of water molecule

DIRECTIONS: With younger children, it will be necessary to introduce the idea of models. Say that in some ways, making human models with toothpicks and clay is like making molecular models with toothpicks and clay. Show children a human model constructed from toothpicks and clay:

FIGURE 5-51. Human model.

Ask children how the model is like a real person (it has two arms, two legs, a body, and a head; the arms, legs, and head are attached in the appropriate places). Ask how the model is different from a real person (it's too small, the colors are wrong, the materials are wrong, it can't move, it isn't alive, it doesn't have any internal organs, etc.). Explain that just as we can construct a human model, we can construct a molecular model, but that just as the human model differs in many ways from a real person, so the molecular model differs in many ways from a real molecule.

Make a model of a water molecule as an example:

FIGURE 5-52. Water (H_2O).

Toothpicks broken in half represent bonds, and clay spheres represent atoms. Let the diameter of the oxygen atom equal approximately 1 in. (2.5 cm) and the diameter of the hydrogen atoms equal approximately ½ in. (1.25 cm).

Ask children how they think the model is like a real water molecule (it has one oxygen atom and two hydrogen atoms, attached in the right places). Ask them how they think it's different (it's much too big, the colors are wrong, the atoms lack internal structure, etc.).

Designate one color for oxygen, one for hydrogen, and one for carbon, and have the children construct models of water, oxygen, carbon dioxide, and hydrogen peroxide molecules (you can either give the children copies of the figures or draw them on the chalkboard):

FIGURE 5-53. Oxygen (O_2).

FIGURE 5-54. Carbon dioxide (CO_2).

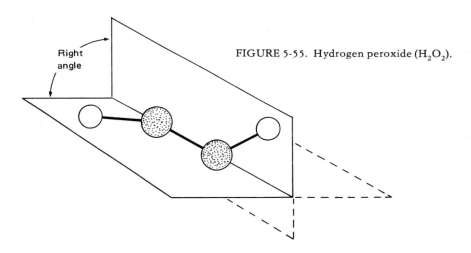

Right angle

FIGURE 5-55. Hydrogen peroxide (H_2O_2).

To construct hydrogen peroxide, the children will need to understand the concept of "right angle." Explain right angles by a drawing and by holding two objects, such as pencils or rulers, perpendicular to each other. Let children find other examples of objects that form right angles: a wall and a floor or ceiling, the spine and cover of a book, the faces of a box, and so on.

After children have constructed the models, discuss each molecule. Ask why they think two toothpicks are used between the atoms in oxygen and carbon dioxide (because two bonds hold the atoms together). Point out that both water and carbon dioxide have three atoms. Ask how these molecules are different from each other (water has one oxygen atom and two hydrogen atoms, whereas carbon dioxide has one carbon atom and two oxygen atoms; water is bent, whereas carbon dioxide is straight; water has single bonds between the atoms, whereas carbon dioxide has double bonds).

MATERIALS: Wooden toothpicks, modeling clay (3 different colors), clay model of methane molecule

DIRECTIONS: Make a methane model as an example:

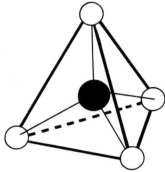

FIGURE 5-56. Methane (CH_4).

Explain that this molecule forms a triangular pyramid with a carbon atom at the center and a hydrogen atom at each corner, and that whenever a carbon atom bonds with four other atoms, the bonds form a triangular pyramid.

Give the children copies of the following figures, or draw them on the chalkboard. Let children identify the pyramids in these figures. (Methanol and acetic acid have one pyramid each; ethanol has two.)

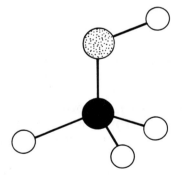

FIGURE 5-57. Methanol (CH_3OH). Wood alcohol.

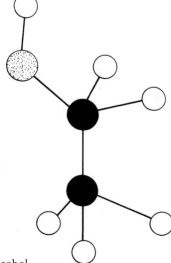

FIGURE 5-58. Ethanol (CH_3CH_2OH). Grain alcohol.

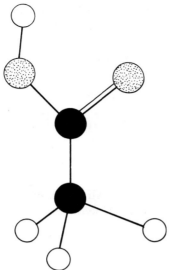

FIGURE 5-59. Acetic acid (CH₃COOH).

Have children construct models of methane, methanol, ethanol, and acetic acid. After they have done this, ask them to find similarities and differences among these molecules.

As a follow-up, have children do research to find out about the molecules they have constructed. From encyclopedias, dictionaries, and science books, they should be able to obtain the following kinds of information:

Water (H_2O). In addition to forming lakes, rivers, oceans, and rain and snow, water is the most abundant substance in the bodies of plants and animals. The human body is over two-thirds water. We know water best in its liquid form that flows and pours, but it can also be a gas or solid. As a gas (steam), it cannot always be felt or seen. Ice is solid water.

Oxygen (O_2), a gas that makes up about one fifth of the Earth's atmosphere, is produced by green plants in the process called photosynthesis. All animals breathe oxygen. Plants also need oxygen to survive. Whenever anything burns, it combines with oxygen.

Carbon dioxide (CO_2) is a gas found in small amounts in the atmosphere. Animals give off carbon dioxide when they breathe. Although plants also give off carbon dioxide, they consume more than they give off, because they use it in photosynthesis to make sugar. Because carbon dioxide does not support fire, it is used in some fire extinguishers.

Hydrogen peroxide (H_2O_2) is a colorless liquid that is used in dilute solutions as an antiseptic and bleach. It tends to break up, forming oxygen and water. Concentrated hydrogen peroxide solutions, which break up rapidly enough to explode, can be used to propel rockets.

Methane (CH_4), a component of natural fuel gas, is produced when plants decay in places where there is little oxygen. Forming explosive mixtures with oxygen or air, it is responsible for mine explosions.

Wood alcohol (CH_3OH) is a colorless liquid made by heating hardwood in the absence of air. Poisonous to humans, it can cause blindness or death when taken into the body. It is used as an antifreeze, a preservative for laboratory specimens, and a raw material for manufacturing other chemicals.

Grain alcohol (CH_3CH_2OH), found in alcoholic beverages, is produced when certain bacteria or yeasts break down sugar or starch in the absence of oxygen. This process is called fermentation. In industry, grain alcohol is sometimes made by synthetic chemical processes. It has many industrial and commercial uses.

Acetic acid (CH_3COOH), the substance that makes vinegar sour, forms when oxygen comes into contact with dilute solutions of grain alcohol. At home, it is used in cooking and pickling; in industry, it is used to manufacture other chemicals.

If you have the background to teach chemical equations, you can have the children use models to balance equations. Have them build the reacting molecules and disassemble them to make the products. They must use all of the reactant atoms— no more, no less. How many of each product molecule can they make?

6 Working with Numbers

Learning the procedures for adding, subtracting, multiplying, and dividing numbers was until recent years the main focus of the elementary school arithmetic program. Often computation was learned by memorization, with little understanding of why procedures work. Since the late 1960s, most arithmetic programs have become *mathematics* programs, of which computation is just one facet. In *An Agenda for Action*,[1] it is recommended that the basic skills in mathematics be defined to encompass more than computational facility, with appropriate computational skills listed as one out of ten skill areas. In our activities, we have tried to avoid rote, routine computation drill and instead to incorporate:

1. understanding,

2. applications, and

3. problem solving.

---

[1]*An Agenda for Action: Recommendations for School Mathematics of the 1980s* (Reston, Virginia: The National Council of Teachers of Mathematics, Inc., 1980), pp. 5–8.

Even young children can and should learn about numbers in a more meaningful way. The Number Charts activities emphasize the order of numbers; patterns of numbers emerge from the Hundreds Puzzle. Understanding of two-digit numbers is stressed in Number Models. Counting by rote is not enough; children need to know that the 2 in 28 has a greater value than the 8. Visual models, such as those provided in these activities, help children reach that understanding.

Four-in-a-Row and Rectangle Multiplication give children the opportunity to learn the basic addition and multiplication facts in an enjoyable, game-playing context. Instead of memorizing isolated facts, children can fit the facts into an addition or multiplication chart in an orderly way. Rectangles are a good visual model for multiplication. This is the basis of the Rectangle Multiplication game, which also utilizes strategy and spatial visualization skills.

The focus of Card Scramble is flexibility with numbers and the four operations. Trial-and-error approaches, rarely used in teaching computation, are encouraged in this activity.

Factual information is used in Exploring Careers, Figuring Salaries, Fields of Study, and Facts About Food. These problems cannot be solved by rote methods; students must think about the situations. These problems provide an opportunity to use computation skills in solving relevant problems.

All of these activities present using numbers in a non-threatening way and allow the children to do creative, open-ended thinking. Most of the problems can be solved in more than one way; there are decisions to be made and strategies to be used. Each child can be successful at her own level.

Number Charts

SKILL AREA: Understanding Numbers

GRADE LEVEL: Primary

STRATEGIES: Using manipulatives
Success for each child
Independent work
Cooperative work

MATH CONCEPTS: Counting, order of numbers

Missing Numbers[2]

MATERIALS: The playing board is a chart with numbers to 50 (or 100) with some numbers omitted (graph paper with 1-in. or 2cm

---

[2]This activity was adapted from Carol Langbort, "Easy Two-Digit Number Activities," *California Mathematics,* 4:2 (October 1979), 12–16.

1			4		6	7		9	
11	12	13		15				19	20
21	22			25	26				30
31		33				37	38	39	
	42		44	45	46				50

FIGURE 6-1. Playing board.

2	3	5	8	10	14	16	17	18	23
24	27	28	29	32	34	35	36	40	41
		43	47	48	49				

FIGURE 6-2. Cut out these individual squares to be placed on the playing board.

squares would work); loose squares with the missing numbers written on them

DIRECTIONS: The child needs a playing board and the loose squares. The task is to place the missing numerals in their correct spaces. Children approach this activity in two distinct ways. They may look for the missing numerals in the order that they are missing. For example, on the sample playing board, the child would first look for the loose 2, then the 3, then the 5, and so on. More advanced children may pick any numeral and find the place where it belongs. For example, a child may take the 35 and place it in the correct space, even though the smaller missing numbers have not yet been found. Two children may play this together, taking turns choosing a number and placing it correctly on the playing board.

Hundreds Puzzle

MATERIALS: Graph paper, 10 units by 10 units (1-in. or 2cm squares); pencil; scissors; paste; construction paper (also 10 by 10 units)

DIRECTIONS: Direct the child to make a hundreds chart number puzzle in the following way. First, number the squares on the graph paper in order from 1 to 100. The numbers may be written either across the rows or down the columns. Paste the chart onto the construction paper. Let this dry overnight. Then cut the chart into 6 or 7 different-shaped pieces by cutting on the lines of the graph paper. Shuffle the pieces. Finally, put the puzzle pieces back together again to form the hundreds chart.

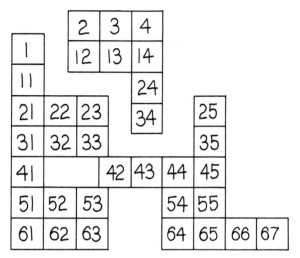

FIGURE 6-3.

Number Models[3]

SKILL AREA: Understanding Numbers

GRADE LEVEL: Primary

STRATEGIES: Using manipulatives
Success for each child
Independent work

MATH CONCEPTS: Counting, order of numbers, place value

Tens and Ones Cards

MATERIALS: 20 cardboard rectangles (5 in. by 8 in. or 12cm by 20cm), glue, scissors, pen, graph paper (½-in. or 1cm squares) cut into single squares and strips of 10 squares

Use these materials to make a set of cards like those on the following page.

---

[3] *Ibid.*

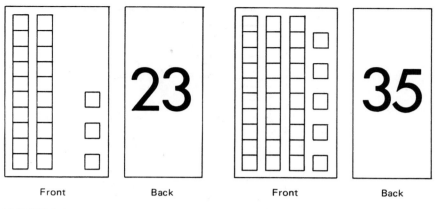

Front	Back	Front	Back

FIGURE 6-4.

Choose any numbers between 11 and 99 for your set.

DIRECTIONS: 1. Stack the cards with the graph paper sides facing up. Have the child count the squares, write the numeral on the paper, and then turn over the card to check the answer. Children can count the squares either by counting the tens and then the ones or one by one.

    2. Stack the cards with the numeral sides facing up. Have the child look at the numeral, then build the model of that number with graph paper that has been cut into strips of ten and loose ones. Finally, the child turns over the card to check that the model matches the picture.

    3. This is a two-person activity. Each child picks a card, numeral side up, then builds a graph paper model. The children each line up their tens and ones in a single column to determine whose number is greater.

*Making Number
Models*

MATERIALS: Strips of tens (1-in. or 2cm graph paper), single 1 in. or 2cm squares, cards (2 in. by 3 in. or 5cm by 8cm) with two-digit numerals on them, paste, construction paper, pencils

DIRECTIONS: The child chooses a card with a number on it, finds enough tens and ones to match this number, and pastes them on construction paper. She then numbers each square. The last number on the square should match the number on the card. Here is a sample of the completed activity on the following page.

FIGURE 6-5.

It is not always obvious to children that two tens equal twenty, even though they may be very good at rote counting by tens. A balance is needed between activities that are essentially memorization, such as counting by tens, and activities such as these that give some meaning to these numbers.

Four-in-a-Row

SKILL AREA: Understanding Numbers

GRADE LEVEL: Primary, intermediate

STRATEGIES: Using manipulatives
Game playing
Many right answers
Many approaches

MATH CONCEPTS: The basic facts of addition and multiplication can be organized on a chart.

5 + 3 and 3 + 5 give the same *sum;* this is an illustration of the *commutative property* of addition.

5 × 3 and 3 × 5 give the same *product;* this is an illustration of the *commutative property* of multiplication.

MATERIALS: Use 1-in. or 2cm graph paper to make a blank addition chart or multiplication chart. This is the gameboard used by both players. (See the sample board under Directions.) Two dice, a different colored crayon for each player. If you wish to practice sums to 18 or products to 81, make your own dice and label them (and the chart) 4, 5, 6, 7, 8, 9.

DIRECTIONS: Each player rolls the dice in turn and records the sum on the addition chart with a colored crayon. Example: if 5 and 3 show on the dice, the 8 may be written in the third row, fifth column or in the fifth row, third column. On each turn a player may write only one number. She must decide which of the two squares will help her to win the game. The first child to write four numbers in a row either horizontally, vertically, or diagonally is the winner.

As a variation, the player may write the sum in *any* of the squares where that sum is appropriate. For example, if 5 and 3 are the numbers rolled, the 8 may be written in any one of these squares: 4 + 4, 6 + 2, 3 + 5, 5 + 3.

+	1	2	3	4	5	6
1						
2						
3					8	
4						
5			8			
6						

FIGURE 6-6.

This game can also be used to practice multiplication. If 3 and 4 are rolled on the dice, the child could write 12 in the third column, fourth row or the fourth column, third row. In the variation described above, she could also include these possible squares: 2 × 6 and 6 × 2.

Rectangle Multiplication

SKILL AREA: Understanding Numbers

GRADE LEVEL: Primary, intermediate

STRATEGIES: Many right answers
Using manipulatives
Success for each child
Game playing

MATH CONCEPTS: A multiplication problem can be visually represented by a rectangle.

5×3 and 3×5 give the same answer and also are the same shaped rectangle; this is an illustration of the *commutative property* of multiplication.

A *factor* of a number divides evenly into that number; a number can have several factors. The factors of 12 are 1, 2, 3, 4, 6, and 12.

MATERIALS: The gameboard, used by both players, is a blank piece of graph paper, at least 16 by 16 units. Each player needs a different colored crayon. You can make the following set of cards:

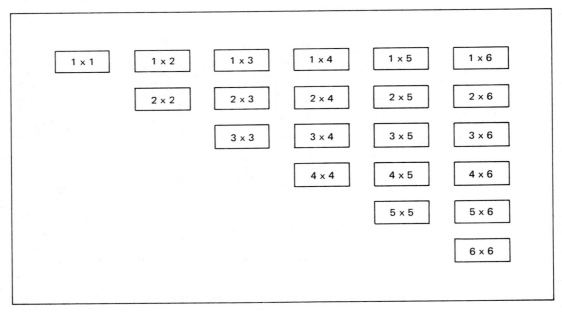

FIGURE 6-7.

DIRECTIONS: Two players take turns. On each turn, a player picks one of the cards, which have been placed face down. Next, she outlines a rectangle on the playing board to represent the fact on her card. (See sample playing board.) Finally, she counts the squares in the rectangle and writes that number inside it. For example, if 2×6 is written on the card chosen, the child can color in a rectangle that is 2 columns and 6 rows or 6 columns

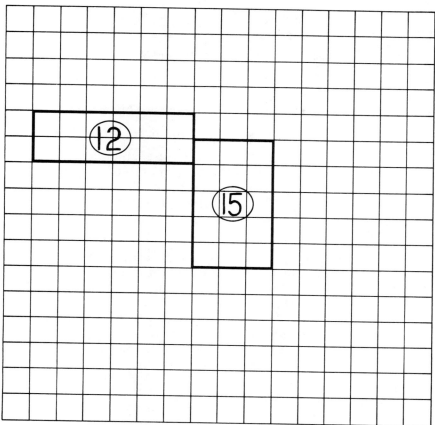

FIGURE 6-8. In this example, one player has marked a
2-by-6 rectangle; the next player has marked a 5-by-3
rectangle.

and 2 rows. This rectangle may be placed anywhere on the
gameboard. The card is not reused.

The second player chooses a card and outlines a rectangle
on the playing board. At least one side of one square must touch
any rectangle that is already drawn on the board. The first
player that is unable to fit a rectangle on the board loses the
game.

A slight variation can give an added dimension to this
game. Make your set of cards with the following *products* on
them instead of *factors*:

1, 2, 3, 4, 5, 6, 8, 9, 10, 12, 15, 16, 18, 20, 24, 25, 36

The game proceeds in a similar manner to the one described.
This time, when a player chooses a card, there is a choice of
rectangles which can be drawn. If the number 8 is chosen, any
one of the following rectangles may be colored in on the

gameboard: 8×1, 1×8, 2×4, 4×2. In this version, the child can think of several pairs of factors and choose a rectangle that will fit on the board.

Card Scramble

SKILL AREA: Understanding Numbers

GRADE LEVEL: Iintermediate, junior high

STRATEGIES: Success for each child
Many right answers
Many approaches
Guessing and testing
Using manipulatives
Independent work

MATH CONCEPTS: The four operations, $+, -, \times, \div$, can be used to combine the numbers on the cards.

Parentheses are used to clarify the grouping of the numbers.

The way the cards are put together can be written symbolically in number sentences.

The *order of operations* determines that multiplication and division are performed before addition and subtraction. Example: $4 + 3 \times 2 - 1 = 9$. 3 was multiplied by 2 before adding 4 and then subtracting 1. With parentheses the answer could be different: $(4 + 3) \times 2 - 1 = 13$.

MATERIALS: A regular deck of playing cards with the picture cards removed, pencils

DIRECTIONS: The child takes the top 5 cards from the shuffled stack. The sixth card becomes the *goal* card. The object of the game is to combine the 5 cards in any way, using any mathematical operation, to form number sentences which equal the goal card. Each of the 5 cards may be used only once in any number sentence.

For example, 7, 10, 5, 8, and 3 are the 5 cards chosen, and the goal card is 9. The object is to arrange the cards in as many different ways as possible so that they equal 9. The score for each number sentence is based on the number of cards used. If all 5 cards are used in any one number sentence, the player receives a bonus of 10 additional points. The number sentences and scores are written on a score sheet.

Here are some number sentences for the cards.

10 minus 7 is 3; 3 times 3 makes 9. Symbolically, this is written as $(10 - 7) \times 3 = 9$. The score for this number sentence is 3, since 3 cards—10, 7, and 3—were used.

8 minus 7 is 1; 1 plus 3 is 4; 4 plus 5 is 9. Using symbols, this is written as $8 - 7 + 3 + 5 = 9$. The score is 4 points, because 4 cards—8, 7, 3, and 5—were used.

10 minus 5 is 5; 5 plus 3 is 8; 8 plus 8 is 16; 16 minus 7 is 9. This is written as $10 - 5 + 3 + 8 - 7 = 9$. The score is 5, plus 10 bonus points, since all 5 cards were used.

This game can be played in a variety of social arrangements. Each child can work independently with her own set of cards. Alternatively, children can work in cooperative groups to make as many different number sentences as possible. Or, this can be an ongoing whole-class activity, where children are continually adding sentences to a classroom chart for particular cards.

Every child has a chance to be successful at her own level. Even the slowest child can find at least one way of putting the cards together to get the number on the goal card. The children are challenged at whatever level they can work comfortably. The game, in fact, encourages children to think more creatively with facts that they already know. Writing the number sentences mathematically and using parentheses appropriately are useful skills reinforced by this activity.

Score Sheet for Card Scramble

Pick 5 cards and 1 goal card. Write them on the blanks below.

_____ _____ _____ _____ _____

GOAL CARD _____

Write 5 number sentences. For each sentence, record how many cards you used. Add a bonus of 10 points each time you use all 5 cards in a sentence. Then total your score.

Number Sentences	How Many Cards Did You Use?	Bonus Points
_____	_____	_____
_____	_____	_____
_____	_____	_____
_____	_____	_____
_____	_____	_____

Total Score: _____ + _____

= _____

Figuring Salaries

SKILL AREA: Applications and Problem Solving

GRADE LEVEL: Intermediate, junior high

STRATEGIES: Content relevance
Estimating
Independent work
Many approaches
Many answers

MATH CONCEPTS: Computation with whole numbers, money, and decimals; measurement conversion

MATERIALS: Each child needs a copy of the worksheet that accompanies this activity, pencil and paper, and a calculator, if available.

DIRECTIONS: The worksheet gives information about the beginning salary offers to women receiving their bachelor's degrees in a recent year.[4] (You should fill in the current minimum wage for comparison.) All of these fields require at least 3 years of high school math and some math in college.

Two children can work together to fill in the worksheet by using the following information:

1. There are 40 hours in a workweek.

2. There are 4.33 weeks in a month.

3. There are 12 months in a year.

You might discuss why there are 4.33 weeks in a month. One child first estimates the answers in her head; then the other figures them out using a calculator or pencil and paper. The children should take turns being the estimator and the calculator user.

After the worksheet is completed, have children answer the following questions:

Which field pays the most? What is the annual salary?

Which field pays the least? What is the annual salary?

How much more money would you earn in one year if you worked in the highest-paying field rather than the lowest-paying field?

How much more would you earn each month?

How much more would you earn each week?

---

[4]The information for this worksheet was compiled from 1978–79 salaries for women reported in *Science Education Databook* (Washington, D.C.: The National Science Foundation, n.d.), p. 138.

Table 6–1: Worksheet for Figuring Salaries

Field	Hourly rate (estimate/exact)	Weekly rate (estimate/exact)	Montbly rate (estimate/exact)	Annual rate (estimate/exact)
Business	********/$6.34	_____/_____	_____/_____	_____/_____
Economics	_____/_____	********/$254	_____/_____	_____/_____
Engineering	_____/_____	_____/_____	********/$1,557	_____/_____
Biological sciences	_____/_____	_____/_____	_____/_____	********/$11,700
Chemistry	_____/_____	_____/_____	********/$1,319	_____/_____
Computer science	_____/_____	********/$318.94	_____/_____	_____/_____
Mathematics	********/$7.53	_____/_____	_____/_____	_____/_____
Minimum wage	********/_____	_____/_____	_____/_____	_____/_____

Encourage children to make up their own questions, using the information on the completed worksheets. They might also find it useful to find out salaries for other fields and add the information to their worksheets.

Exploring Careers[5]

SKILL AREA: Applications and Problem Solving

GRADE LEVEL: Intermediate, junior high

STRATEGIES: Many approaches
          Independent work
          Estimating
          Content relevance
          Modeling new options

MATH CONCEPTS: Computation, measurement conversion

MATERIALS: Pencils, paper

DIRECTIONS: Give each student a copy of the following problem. Have her read the problem, and give her an opportunity to ask questions before she solves it.

> High school students Betsy, Martha, and Lynn are discussing their future careers.
>
> Lynn says, "I can expect to earn $1,795 per month when I graduate from college if I major in petroleum engineering."
>
> Betsy says, "If I major in social science, I can expect to earn approximately one half the average starting salary of a petroleum engineer when I get out of college."
>
> Martha says, "If I enter a carpentry apprenticeship program, I'll be earning $13.00 an hour by the time you graduate from college."
>
> Assuming the young women go ahead with their plans, what will be the annual salaries of each of them the year that Betsy and Lynn graduate from college? (Assume that are 40 hours in a workweek and 50 weeks in work year.)

Your students may be unfamiliar with petroleum engineering and the fields of social science. Social science includes the following areas of study: political science, sociology, psychology, economics, and anthropology. Ask your students what social scientists do.

---

[5]This activity was contributed by Sherry Fraser, math specialist, Math/Science Desegregation Project, Novato (California) Unified School District.

A petroleum engineer's main concern is with improving the production of oil or gas wells. They improve well designs and advise on the maintenance of wells. To determine where to locate wells, they study earth samples or even conduct geological surveys. Their job also includes keeping production records on wells and preparing regular engineering reports.

Discussion Questions:

You may increase awareness of careers requiring math by exploring these questions:

1. *How much math is required for each of these occupations?* Four years of high school math are strongly recommended to enter a program in petroleum engineering. An additional three and a half years of math are required during college. Two years of high school math are required to enter a social science program at the university. At least one college math course must be taken before graduation. Graduate study in social sciences requires additional math, including statistics. Math through geometry is recommended to enter a carpentry apprenticeship program. Applied math constitutes a large portion of the weekly class requirement during the four years of apprentice training.

2. *What are the possibilities of finding a job in these occupational fields?* The petroleum engineer will probably have many job offers. The social science major will have to find a way to apply her skills to the job market. She will face stiff competition for most of the jobs available for people with her educational background. The carpenter will have little difficulty finding a job, but work may be short-term or seasonal.

3. *What type of salary can be expected with 20 years' experience in the field?* The petroleum engineer can expect advancement in her salary and career as her experience increases. In the social sciences, intense competition for the few high level positions available makes salary advances less predictable. Carpenters often stay at journey wages, which are renegotiated periodically.

4. *What are some of the other occupational characteristics of the young women's career choices?* For example, what will the working conditions be like? Will she work alone or with others? Will she have direct supervision or be responsible for her own work? Will she work an eight-hour day, five days a week, or will the hours be flexible? What type of benefits come with the job? (Students can talk to people and look in books to find the answers to these questions.)

Solution:

Lynn (petroleum engineer)

$1795 per month
× 12 months per year

3590
1795

21540 $21,540 a year

Betsy (social science)

10,770

2 / $21,540 $10,770 a year

Martha (carpenter)

$13.00 per hour
× 40 hours per week

$520.00 per week
× 50 weeks per year $26,000 a year

$26,000.00

Fields of Study

SKILL AREA: Applications and Problem Solving

GRADE LEVEL: Junior high

STRATEGIES: Content relevance
Estimating
Many approaches
Many answers
Independent work

MATH CONCEPTS: Interpreting information on a chart can help you solve math problems.

Using *percentages* is a valuable way of describing relationships.

Estimating is an effective procedure for solving problems with large numbers.

Rounding off numbers helps in estimating.

MATERIALS: A copy of Table 6–2 and the worksheet that accompanies this activity for each student

DIRECTIONS: Have students answer these questions by using the information in Table 6–2. They should not use pencil and paper or calculators. Encourage them to do all the thinking in their heads. They should round off the numbers so that they can estimate the answers more easily.

1. In which of these fields were most degrees given?

2. Which field was the next most popular?

3. In which field is there *about half as many women as men* receiving degrees?

4. In which two fields are there *about three times as many men as women* who received degrees?

5. In which two fields are there *about three times as many women as men* who received degrees?

6. In which two fields was the total number of degrees given less than fifteen thousand?

In the discussion following this work, ask for the different procedures students used to estimate their answers.

The questions listed here are meant to be examples. Many more problems can be made up using the given information. Students should be encouraged to make up their own problems.

Table 6–2*: Number of Men and Women
Receiving Degrees in Various Fields

Field of study	*Men*	*Women*
Biological sciences	34,474	19, 719
Business and management	117,510	36,252
Computer and information sciences	4,887	1,539
Education	40,410	104,988
Engineering	47,437	2,240
Foreign languages	3,410	10,892
Psychology	20,692	27,102

*Source: "Degrees Awarded in 1977," *Scientific, Engineering, Technical Manpower Comments*, vol. 16, no. 5 (June 1979).

Filling in the blanks in part one of the worksheet gives students a chance to estimate totals and percentages. Again, instruct them to do this work in their heads and to round off the numbers so that they can estimate more easily. First, they should write their estimated totals. Then they should estimate the percentage of men and women receiving degrees in each field. Remind them that their percentages should add up to 100%.

Let students use calculators or paper and pencil to fill in part two of the worksheet. Ask them to compare the exact answers with their estimated answers. They might be surprised by how close they came to the exact answer.

Table 6–3. Worksheet for Fields of Study

| Field of Study | Part I: Estimated | | | Part II: Exact | | |
	Total	% Women	% Men	Total	% Women	% Men
Biological sciences	___	___	___	___	___	___
Business & management	___	___	___	___	___	___
Computer and information sciences	___	___	___	___	___	___
Education	___	___	___	___	___	___
Engineering	___	___	___	___	___	___
Foreign languages	___	___	___	___	___	___
Psychology	___	___	___	___	___	___

Students might find it enlightening to research current salaries for people starting out in the listed fields. They can then make two lists, one ranking the fields according to the percentage of women receiving degrees, the other ranking the fields according to starting salaries. Discuss any similarities between the two lists.

Facts About Food

SKILL AREA: Applications and Problem Solving

GRADE LEVEL: Junior high

STRATEGIES: Many approaches
Many right answers
Estimating
Cooperative work
Content relevance

MATH CONCEPTS: Computation using fractions, decimals, and percentages; measurement conversion

MATERIALS: Copies of the problems for each student; calculators, if available; pencil and paper

DIRECTIONS: The following problems contain factual information.[6] There are both computation problems and discussion questions based on this information. Have the students think about the problems and estimate their answers before doing the computations. If possible, have them use calculators to find the answers. Working in groups is an effective way for students to solve problems of this type. Encourage discussion of the different methods they used to arrive at their answers. They should also share their various ideas on the discussion questions.

1. In 1976 the average American ate 4.4 pounds of butter. That is 1/4 of the amount of butter that was eaten in 1910. About how much butter did the average person eat in 1910? Why do you think we eat less butter now?

2. In 1976 the average American ate 95.6 pounds of beef. How much beef is this per month? How much beef is this per day? In 1976 we ate 90% more beef than in 1950. About how much beef did the average American eat in 1950?

3. In 1976 each person drank 92.5 quarts of milk. How many 8 ounce glasses is that per day?

---

[6]Mike Feinsilber and William B. Mead, *American Averages* (Garden City, N.Y.: Doubleday & Co., Inc., 1980), pp. 283–85.

4. In 1976 the average American ate 15.9 pounds of cheese. That is 50% more cheese than was eaten in 1969. How much cheese did the average American eat in 1969? Why do you think cheese is so much more popular now?

5. In 1976 the average American ate 17 pounds of candy. That is 1/5 less than in 1968. How much candy did the average American eat in 1968? Why do you think we are eating less candy now?

6. Is the "average American" a real person? What does this term mean? Does your diet differ from this average? What other people in our population would have a very different diet? (Diet can vary with income, ethnic group, age, size, etc.).

7. How can you find out what the average student in your class eats?

7 Logical Reasoning

SCIENCE ACTIVITIES

Logical reasoning consists of many different subskills, all of which are important in science and math. Jean Piaget and others have shown that logical reasoning develops gradually and that children need to learn how to reason using concrete objects before they can reason abstractly. The hands-on activities in this chapter help to build the necessary foundation for the following types of logical reasoning:

1. sorting and classifying,
2. deductive reasoning,
3. combinatorial reasoning,
4. controlling variables, and
5. probabilistic reasoning.

Sets and Games with Rocks, Seeds, and Leaves help children develop sorting and classifying skills. Classification is important in both the biological and physical sciences. Scientists have classification systems for objects such as plants, animals, and minerals. In the activity, however, children do not learn scientific classification systems; this would be a memorization task that they could perform with various degrees of accuracy without understanding its logical basis. Instead, using simple properties such as size, texture, color, and shape, they develop their own systems to sort and classify rocks, seeds, or leaves.

The guessing games with rocks, seeds, and leaves help children learn to use deductive reasoning. In these games, players ask questions and draw conclusions based on the answers they receive. What Makes the Hoppits Hop? and Kitchen Chemistry also give children experience with deductive reasoning.

Combinatorial reasoning is the ability to identify all possible combinations of objects or events. In Pea Plants, children identify various combinations of pea plant characteristics, such as seed shape, flower color, and plant height. In Kitchen Chemistry, they mix various combinations of water, white vinegar, baking soda in water, and water in which red cabbage has been boiled. Their task is to make a pink substance, a green substance, and a substance that fizzes.

Control of variables is an essential skill in designing and implementing scientific experiments. In controlling variables, a scientist changes only one factor at a time. For example, to determine the effects of a particular mineral on plant growth, a scientist would have to raise two groups of plants under identical conditions, except that only one group would receive the mineral. If the groups were also exposed to different amounts of light, the scientist would not be able to determine whether the mineral or the light was responsible for differences in growth. Which Material Is Strongest? helps children learn to control variables. This important skill is also emphasized in some of the activities in chapter eight.

Probabilistic reasoning requires determining the relative frequency with which an event will occur. Suppose that you are flipping two coins. What is the probability that two heads will come up? Two tails? One head and one tail? Since there are four distinct ways that the coins can land, the probability of flipping two heads is $\frac{1}{4}$, as is the probability of flipping two tails. The probability of flipping one head and one tail, however, is $\frac{1}{2}$, because there are two ways to make this combination.

Inheriting Traits, the final activity in this section, helps children develop probabilistic reasoning. This activity introduces some basic concepts from genetics, the branch of biology that deals with heredity. In addition to being fundamental to genetics, probabilistic reasoning is important in all fields of science, because statistical tests are used to analyze data, and conclusions are drawn on the basis of the *probability* that an observed event was due to chance.

Sets and Games with Rocks, Seeds, and Leaves

To do these activities you will need 10 to 20 rocks, seeds, or leaves for each pair of children. If seeds or leaves are used, they should come from 10 to 20 different kinds of plants. It is okay, however, to use rocks of the same kind. This is because rocks of the same kind can be distinguished by size or shape, whereas seeds and leaves of the same kind are often identical. Children gain more from the sorting activity, Making Sets, if they have to group objects together that are *not* identical. In the Guessing Games it is necessary that each object have unique properties.

Making Sets

SKILL AREA: Logical Reasoning (Sorting and Classifying)

GRADE LEVEL: Primary

STRATEGIES: Using manipulatives
              Success for each child
              Many right answers
              Many approaches
              Cooperative work

SCIENCE CONCEPT: Objects found in nature can be sorted and classified according to easily recognizable properties, such as size, texture, color, and shape.

MATERIALS: 10 to 20 rocks, seeds, or leaves for each pair of children; 8½-in.-by-11 in. (or 20 cm by 30 cm) graph paper with 1-in. (or 2 cm) squares) (optional); pencils (optional); crayons (optional)

DIRECTIONS: Each pair of children should have 10 to 20 rocks, seeds, or leaves. In a classroom, everyone should work with the same kind of material (rocks, seeds, or leaves). If each child brings 10 rocks, seeds, or leaves to school, there should be plenty of material. At home, you can be your child's partner.

Have children sort their objects into a number of groups. Ask them to put objects together that they think go together, emphasizing that there are many different ways to do this. Some children will stop after sorting the objects into two to four major categories. Others will have subcategories based on additional properties (see Figure 7-1).

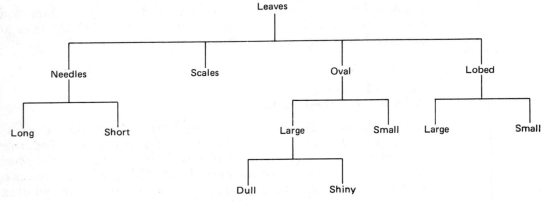

FIGURE 7-1. Leaf classification based on three properties (shape, size, shininess).

Ask the children to explain their systems for sorting. Most children will use size, texture, color, or shape. With rocks it is also possible to use hardness, shininess, patterns, crystals, or weight. Some children might consider vein patterns or shininess of leaves. You can repeat the activity several times, having the children devise different systems for sorting their objects.

Children can make bar graphs to show their systems for sorting. (See chapter five for activities to introduce graphing.) A pictorial graph can be made by marking off a column along the horizontal axis for each category and representing each object by a picture:

FIGURE 7-2. Pictorial bar graph based on seed shape.

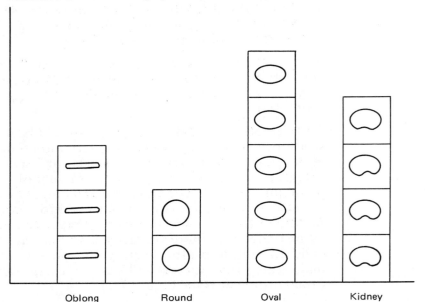

Alternatively, a somewhat more abstract graph can be made by putting numbers on the vertical axis and representing each object by an X:

FIGURE 7-3. Bar graph based on seed shape.

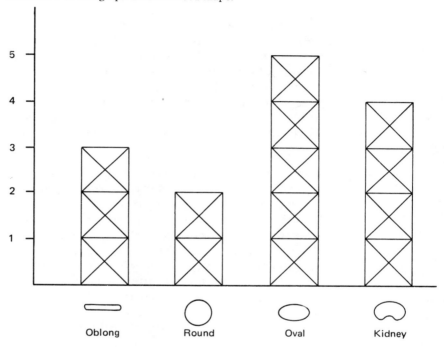

Guessing Games

SKILL AREA: Logical Reasoning (Deductive Reasoning)

GRADE LEVEL: Primary, intermediate

STRATEGIES: Using manipulatives
Success for each child
Many approaches
Guessing and testing
Game playing

SCIENCE CONCEPT: Objects found in nature can be sorted and classified according to easily recognizable properties, such as size, texture, color, and shape.

MATERIALS: 10 to 20 rocks, seeds, or leaves for each pair of children; paper; pencils

DIRECTIONS: *Guess My Set.* This is a two-person game which is played with rocks, seeds, or leaves. Player one chooses a set-determining property, such as size, texture, color, or shape, and writes it on a piece of paper. (Writing helps children clarify their thinking and prevents them from forgetting or changing their set midway through the game.) Player two is not allowed to see what is written on the paper. The object of the game is to figure out, in as few turns as possible, what is written on the paper.

Player two picks objects, one at a time, and asks, "Does this go inside or outside the set?" Player one answers, then puts player two's object into one of two piles: objects inside the set or objects outside the set. Point out to children that objects placed outside the set give just as much information as objects placed inside the set. At any time, player two may guess the set-determining property. If the guess is correct, player two is the winner; if it is incorrect, player one is the winner.

Guess My Object. This is another two-person game which is played with rocks, seeds, or leaves. Player one selects an object and writes a description of it on a slip of paper, which player two is not allowed to see. The description should be brief; for example, "pumpkin seed," "maple leaf," or "smooth gray egg-shaped rock." The object of the game is for player two to identify player one's object in as few turns as possible.

Player two can ask questions with "yes" or "no" answers, such as "Is it round?" "Is it rough?" or "Is it shiny?" According to player one's answers, objects that can no longer be considered are removed from the pile. At any time, player two may guess which object player one is thinking of. A correct guess makes player two the winner; an incorrect guess makes player one the winner.

Possible extensions of Guess My Set and Guess My Object:

1. Play with more than one kind of object; for example, both leaves and rocks.

2. Play with more than two people.

3. Play with points.
    a. Player two gets 1 point for each question and 5 points for an incorrect guess.
    b. Play an even number of games, taking turns being player one and player two.
    c. Player with the fewest points at the end is the winner.

What Makes the Hoppits Hop?

SKILL AREA: Logical Reasoning (Deductive Reasoning)

GRADE LEVEL: Primary

STRATEGIES: Using manipulatives
Cooperative work

SCIENCE CONCEPT: Many different factors affect the behavior of animals. Two very important factors are food and temperature.

MATERIALS: Red, orange, blue, and green construction paper; tape; scissors

DIRECTIONS: Hoppits are make-believe animals. In this activity, children pretend they are hoppits in order to figure out what makes hoppits hop.

Hoppits can be either hot or cold and can eat either mabbles or prings:

a   b

FIGURE 7-4. Mabble (a) and Pring (b).

Divide children into four groups (at home you will need four people, one to act out each part):

Group 1: *Hot hoppits who have eaten prings.* Give each child in this group a slip of red paper to represent hotness and a piece of green paper to make a pring.

Group 2: *Hot hoppits who have eaten mabbles.* Give each child a slip of red paper to represent hotness and a piece of orange paper to make a mabble.

Group 3: *Cold hoppits who have eaten prings.* Give each child a slip of blue paper to represent coldness and a piece of green paper to make a pring.

Group 4: *Cold hoppits who have eaten mabbles.* Give each child a slip of blue paper to represent coldness and a piece of orange paper to make a mabble.

Have the children tape both their slips of paper showing hotness or coldness and their mabbles or prings to their clothing.

Tell children that first they will be *tree hoppits*. Speak with the groups individually to tell them what to do when you say "Go":

Group 1: Hop.

Group 2: Hop.

Group 3: Move in any way other than hopping.

Group 4: Move in any way other than hopping.

Send one group to each corner of the room. Give the signal "Go" to one group at a time, so that the other groups can watch what happens.

Discuss the following propositions:

1. Hotness makes hoppits hop.

2. Prings make hoppits hop.

3. Coldness makes hoppits hop.

4. Mabbles make hoppits hop.

Ask children whether each statement is true or false. (1 is true; 2, 3, and 4 are false.) Talk about why 2, 3, and 4 are false. (We know that coldness does not make hoppits hop because none of the cold ones hopped. If mabbles or prings could make the hoppits hop, group three or group four should have hopped.)

Tell children that next they will be *creek hoppits*. Give them the following instructions:

Group 1: Move in any way other than hopping.

Group 2: Hop.

Group 3: Move in any way other than hopping.

Group 4: Hop.

Go through the four propositions again, asking if each one is true or false for creek hoppits. (In this case, 4 is true; 1, 2, and 3 are false.)

Finally, ask the children to be *ground hoppits*. Give them the following instructions:

Group 1: Hop.

Group 2: Hop.

Group 3: Hop.

Group 4: Move in anyway other than hopping.

Review the four propositions once again, asking if each one is true or false. (1 and 2 are true: either hotness or prings will make ground hoppits hop; we know that neither coldness nor mabbles makes them hop, because group four did not hop.)

You can repeat this activity on another occasion, having the children be beach hoppits, desert hoppits, meadow hoppits, forest hoppits, or mountain hoppits. Make up instructions to illustrate various combinations of food and temperature that could make the hoppits hop.

Kitchen Chemistry

SKILL AREA: Logical Reasoning (Deductive Reasoning and Combinatorial Reasoning)

GRADE LEVEL: Primary, intermediate

STRATEGIES: Using manipulatives
Guessing and testing
Independent work
Content relevance

SCIENCE CONCEPT: Some substances *react* chemically with each other to form new substances.

MATERIALS: Cabbage water (instructions for preparing below), white vinegar, baking soda, water, straws, 4 large glasses, small paper cups

DIRECTIONS: Prepare cabbage water by boiling red cabbage leaves, broken into small pieces, for approximately 10 minutes or until the water turns a deep purple. Use approximately one cup of water for each cabbage leaf.

Label the glasses 1 to 4 and fill them as follows:

1. water

2. white vinegar

3. water with baking soda dissolved in it

4. cabbage water

Don't tell the child what the 4 glasses contain until the end of the experiment. The task is to make 3 new substances: a pink substance, a green substance, and a substance that fizzes. In the small paper cups, the child should mix 2 liquids at a time. Ask her to keep a record of her results.

Which 2 liquids made a pink solution? (2 and 4)

Which 2 made a green solution? (3 and 4)

Which 2 made a solution that fizzes? (2 and 3)

Tell the child what the original 4 liquids were, but do not tell her which glass contained which liquid. Give her the following propositions, and let her figure out what was in each glass.

1. When vinegar and cabbage water are mixed, the solution turns pink.

2. When baking soda in water is mixed with cabbage water, the solution turns green.

3. When vinegar and baking soda are mixed, the solution fizzes.

4. Water does not react with any of the other substances.

Discuss the following chemical reactions:

Vinegar contains *acetic acid;* baking soda is sodium bicarbonate, a *base.* The purple cabbage dye turns pink when it reacts with an acid and green when it reacts with a base.

When acetic acid reacts with sodium bicarbonate, one of the products formed is carbon dioxide gas. Release of the gas produces bubbles.

Possible extension: Have the child test other substances with cabbage water, such as lemon juice, salt in water, aspirin in water, and baking powder in water. Ask her whether each substance is acidic, basic, or neutral.

Pea Plants

SKILL AREA: Logical Reasoning (Combinatorial Reasoning)

GRADE LEVEL: Intermediate

STRATEGIES: Using manipulatives
Success for each child
Independent work

SCIENCE CONCEPT: Individuals of a plant or animal species can have different combinations of inherited characteristics.

MATERIALS: Pencil, paper, crayons

DIRECTIONS: People, dogs, and domestic cats provide familiar examples of species having a lot of variation between individuals. Let the child identify some of the variable characteristics of each species (eye color, hair color, size, etc.). Distinguish between characteristics that can be inherited and those that cannot (for example, skin color and clothing color).

Select 2 pea plant characteristics from Table 7–1. Tell the child the 2 forms of each characteristic. For example, if you select pod color and flower color, say that pea plants can have either green or yellow pods and either red or white flowers. Ask the child to tell you the 4 possible combinations of these characteristics (green pods, red flowers; green pods, white flowers; yellow pods, red flowers; yellow pods, white flowers).

Table 7–1. Characteristics of Pea Plants

1.	*Characteristics*	
1.	Seed shape	round, wrinkled
2.	Seed color	yellow, green
3.	Pod shape	narrow, wide
4.	Pod color	yellow, green
5.	Flower color	red, white
6.	Flower position	ends of stem, sides of stem
7.	Plant height	tall, short

Select 2 more characteristics from Table 7–1. Give the child a piece of paper. Have her fold the paper into quarters and open it again. Using the 2 selected characteristics, ask her to draw a different-looking pea plant in each box as shown in Figure 7-5. The drawings can be simple and schematic; showing the different combinations of characteristics is more important than drawing a scientifically accurate pea plant. When only seed and pod characteristics are used, the entire plant need not be drawn.

Continue the activity, using 3 characteristics instead of 2. Because there are 8 combinations of each set of 3 characteristics, the child should fold her papers into eighths.

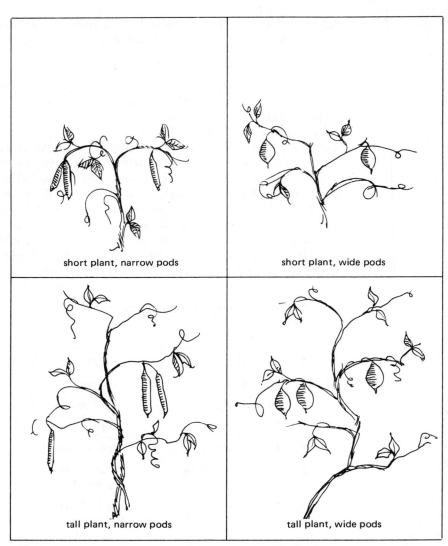

short plant, narrow pods	short plant, wide pods
tall plant, narrow pods	tall plant, wide pods

FIGURE 7-5. Combinations of pod-shape and plant height in pea plants.

Give the child the remaining pea plant characteristics and challenge her to figure out how many different-looking plants you can make with 4, 5, 6, or 7 characteristics. Can she find a rule that will tell her how many combinations there are of *any* number of characteristics? Discuss the answers on another day, after she has had time to think about the problem. Table 7–2 shows that you multiply the number of combinations by 2 each time you add a characteristic.

Table 7–2. Combinations of 1–7 Characteristics, Each of Which Has Two Possible Forms

Number of characteristics	Number of combinations
1	2
2	4
3	8
4	16
5	32
6	64
7	128

Which Material Is Strongest?

This activity should be done after hands-on investigative activities that emphasize making a fair test, that is, controlling variables. See, for example, Plant Growth, Mealworm Behavior, and Paper Airplanes in chapter eight.

SKILL AREA: Logical Reasoning (Controlling Variables)

GRADE LEVEL: Intermediate, junior high

STRATEGIES: Using manipulatives
             Many approaches
             Cooperative work

SCIENCE CONCEPTS: Some scientists and engineers develop new materials and test their strength.

In scientific experiments, one factor should be tested at a time, while all other factors are held constant. This is called fair test, or control of variables.

MATERIALS: A set of 12 materials cards (shown at end of activity) for each group of children, paper, pencils

DIRECTIONS: Children work in groups of 2 to 6 to determine the relative strength of the following make-believe materials: orot, tagib, and gar. To do this, they must simulate and compare the bending of rods that have the same shape and thickness and the same size weight attached.

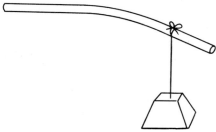

FIGURE 7-6. Rod with weight attached.

The purpose of this activity is to help children learn to make a fair test.

Make enough materials cards so that each group will have one complete set. The cards should be shuffled and divided equally among the members of the group. For groups of 5, eliminate any 2 cards.

Each materials card gives the shape, diameter, and material of a rod and says how much this rod bends when a light or heavy weight is attached to it. You should make the following assumptions:

All the rods are the same length.

All the narrow rods have the same diameter.

All the medium rods have the same diameter.

All the wide rods have the same diameter.

All the light weights are the same.

All the heavy weights are the same.

Children will use their arms to demonstrate the bending. Make sure they understand that the more a rod bends, the weaker it is.

FIGURE 7-7. Degrees of bending.

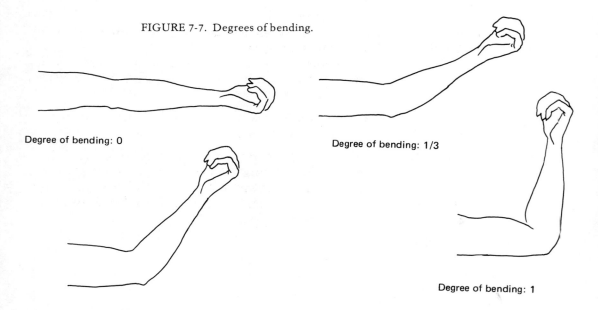

Degree of bending: 0

Degree of bending: 1/3

Degree of bending: 2/3

Degree of bending: 1

Ask children to find out which material is strongest and which is second strongest. The object is to solve the problem as a group, not to see who can get the answers first. These are the rules:

1. Do not let anyone see your cards.

2. You may ask questions about the shape, diameter, and material of the rods specified on another person's cards.

3. You may answer questions about the shape, diameter, and material of the rods specified on your cards.

4. You may ask someone to show you how much a particular rod bends when a light or heavy weight is attached to it.

5. If someone asks you how much a particular rod bends when a light or heavy weight is attached to it, you may answer by demonstrating the bending with your arm.

When everyone has finished, have children explain how they solved the problem. There are several ways to do this. For example, the bending of the rods specified on cards C, G, and K can be compared. It is also possible to look at 2 materials at a time, as, for example, by comparing the bending of the rods specified on cards C and G, then comparing the bending of those specified on A and I. Be sure that the concept of a fair test is brought out: rods of the same shape and diameter should be compared with the same weight attached; only the material should vary. Tagib is the strongest, orot the second strongest.

FIGURE 7-8. Materials Cards.

Do not let anyone see this card.	Do not let anyone see this card.
You may share the following information:	You may share the following information:
Material: orot	Material: orot
Diameter: wide	Diameter: medium
Shape: cylindrical	Shape: cylindrical
You may demonstrate this bending with your arm:	You may demonstrate this bending with your arm:
Light weight: 1/3	Light weight: 2/3
Heavy weight: 2/3 A	Heavy weight: 1 B

FIGURE 7-8 (continued).

Do not let anyone see this card.
You may share the following information:
Material: orot
Diameter: medium
Shape: rectangular
You may demonstrate this bending with your arm:
Light weight: 1/3
Heavy weight: 2/3   C

Do not let anyone see this card.
You may share the following information:
Material: orot
Diameter: extra wide
Shape: rectangular
You may demonstrate this bending with your arm:
Light weight: 0
Heavy weight: 0   D

Do not let anyone see this card.
You may share the following information:
Material: tagib
Diameter: wide
Shape: cylindrical
You may demonstrate this bending with your arm.
Light weight: 0
Heavy weight: 1/3   E

Do not let anyone see this card.
You may share the following information:
Material: tagib
Diameter: narrow
Shape: cylindrical
You may demonstrate this bending with your arm:
Light weight: 1
Heavy weight: 1   F

Do not let anyone see this card.
You may share the following information:
Material: tagib
Diameter: medium
Shape: rectangular
You may demonstrate this bending with your arm:
Light weight: 0
Heavy weight: 1/3   G

Do not let anyone see this card.
You may share the following information:
Material: tagib
Diameter: wide
Shape: rectangular
You may demonstrate this bending with your arm:
Light weight: 0
Heavy weight: 0   H

Do not let anyone see this card.
You may share the following information:
Material: gar
Diameter: wide
Shape: cylindrical
You may demonstrate this bending with your arm:
Light weight: 2/3
Heavy weight: 1   I

Do not let anyone see this card.
You may share the following information:
Material: gar
Diameter: medium
Shape: cylindrical
You may demonstrate this bending with your arm:
Light weight: 1
Heavy weight: 1   J

FIGURE 7-8. (continued).

Do not let anyone see this card.	Do not let anyone see this card.
You may share the following information:	You may share the following information:
Material: gar	Material: gar
Diameter: medium	Diameter: narrow
Shape: rectangular	Shape: rectangular
You may demonstrate this bending with your arm:	You may demonstrate this bending with your arm:
Light weight: 2/3	Light weight: 1
Heavy weight: 1 K	Heavy weight: 1 L

Inheriting Traits

Children will learn about dominant and recessive genes and the role of chance in heredity. These concepts are important in many different life-science fields, including genetics (study of heredity), genetics counseling (counseling of people who may carry genes for abnormalities such as albinism or Tay-Sachs disease), agriculture, and animal husbandry.

SKILL AREA: Logical Reasoning (Combinatorial Reasoning and Probabilistic Reasoning)

GRADE LEVEL: Intermediate, junior high

STRATEGIES: Using manipulatives
            Independent work
            Guessing and testing
            Content relevance

SCIENCE CONCEPTS: Traits that we inherit from our parents, such as hair color, eye color, and height, are determined by *genes*.

   Often, a trait is controlled by two genes, one inherited from our mother and one inherited from our father.

   We need only *one* copy of a *dominant* gene in order to show that trait.

   We need *two* copies of a *recessive* gene in order to show that trait.

   Chance plays an important role in determining the combination of genes we inherit.

MATERIALS: Copies of Table 7–3 and pattern for triangular pyramid, pencils, paper, scissors, tape, small paper bags

DIRECTIONS: Each child should have paper, a pencil, a small paper bag, a copy of Table 7–3, and the following pattern for a triangular pyramid.

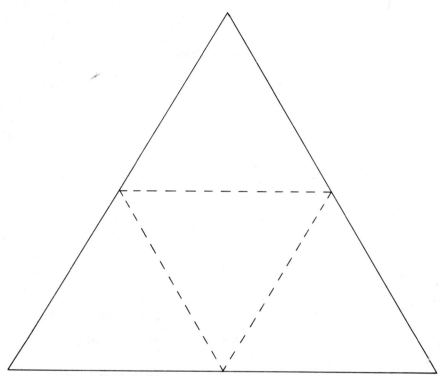

FIGURE 7-9. Triangular pyramid.

Table 7–3. Inheritance of Human Traits

Dominant	Recessive
BROWN EYES	blue eyes
DARK HAIR	blond hair
NONRED HAIR	red hair
CURLY HAIR	straight hair
NIGHT BLINDNESS	normal vision
FARSIGHTEDNESS	normal vision
NORMAL VISION	nearsightedness
BROAD LIPS	thin lips
TONGUE-ROLLING ABILITY	lack of tongue-rolling ability

Each child selects a trait, such as eye color, from Table 7–3. Assuming that a man and a woman, whom we'll call Mr. and Ms. Crandall, each have one dominant gene and one recessive gene

146

for this trait, the child will determine the possible traits of their offspring.

Ask children whether Mr. and Ms. Crandall show the dominant or recessive trait (both show the dominant trait). Make sure that everyone understands the following rules: we need only one copy of a dominant gene in order to show that trait, and we need two copies of a recessive gene in order to show that trait.

There are 4 possible combinations of genes that a son or daughter of Mr. and Ms. Crandall could inherit. Children should find these combinations. Making a chart such as the following will help:

Parent 1

	BROWN	blue
blue (Parent 2)	BROWN blue	blue blue
BROWN (Parent 2)	BROWN BROWN	blue BROWN

FIGURE 7-10.

UPPERCASE letters represent dominant genes; *lowercase* letters represent recessive genes.

Have children write one gene combination on each triangle of the pyramid, cut out the figure, fold it on the dotted lines, and tape the edges together. After predicting whether Mr. and Ms. Crandall's child will show the dominant or recessive trait, they shake the pyramid in the paper bag, then throw it out on a table. The side landing face down determines the child's genes. Have children repeat this procedure several times. They should keep track of the results.

What is the probability that the child will have the recessive trait? (¼) The dominant trait? (¾) Forty throws of the pyramid should produce 10 children with the recessive trait and 30 with the dominant trait. How closely do the children's results match the expected fractions?

Possible extensions: Have children select another trait from Table 7–3. Repeat the activity, starting with either of the following conditions:

1. One parent has two recessive genes; the other parent has one dominant gene and one recessive gene.

2. One parent has two recessive genes; the other parent has two dominant genes.

The activity can also be done with 2 traits. Make a separate triangular pyramid for each trait. Shake and throw the pyramids together many times, keeping track of the results. What is the probability that the child will have one of the recessive traits? Neither recessive trait? Both recessive traits?

MATH ACTIVITIES

The activities in this section exemplify the same logical reasoning skills as the science activities in the first part of this chapter. In Sets and Games with Playing Cards, children sort and classify playing cards according to color, number, and suit; they also practice recognizing numerals and shapes. When playing the guessing games, children use deductive reasoning: during the game process they draw conclusions which help them "guess" answers.

Even young children benefit from playing the Strategy Games, Hex and Nim. Trial-and-error approaches soon lead to the development of winning strategies. As the games progress and more information is obtained, conclusions are based on deductive reasoning.

Solving the Four Scientists problem gives students yet another chance to use deductive reasoning. Combining the given statements in a logical way leads to conclusions that help to solve the problem.

A combination is a selection of things in which the order is not important. Finding all possible combinations is the task in A Shopping Spree. Although there are mathematical formulas for finding the number of different combinations, our emphasis is on having students organize the information so that they know when they have found all the possibilities. Students can also approach Four Scientists by finding all possible combinations.

Sets and Games with Playing Cards

These activities use playing cards to teach the logical thinking skills of sorting and classifying. Children will organize the cards according to their attributes of number, color, and suit. These activities complement Sets and Games with Rocks, Seeds, and Leaves.

SKILL AREA: Logical Reasoning (Sorting and Classifying
and Deductive Reasoning)

GRADE LEVEL: Primary

STRATEGIES: Guessing and testing
Many right answers
Using manipulatives
Success for each child
Independent work
Many approaches
Game playing

MATH CONCEPTS: These activities provide practice in counting
and recognition of numerals and shapes.
You can sort playing cards in many different ways.

MATERIALS: A deck of playing cards; 10 label cards (directions
follow); a tray, such as a shoe box cover
For the following 4 activities you will need 16 cards: 4
numbers in each of the 4 suits. For example, the 2, 3, 4, 5 in all
suits can make up the set.
You will also need to make a set of 10 label cards like these:

FIGURE 7-11.

Matching Cards

DIRECTIONS: Before starting the activity, you might review the
names of the suits and numerals on the cards. One way of doing

this is to turn over a label card such as and ask the child to find all the cards that have this shape on it. Then ask, "What is the name of this shape?" You can check on numeral and color recognition in the same way, using the appropriate cards. With children who cannot yet read, write the "red" card with red ink and the "black" card with black ink.

In Matching Cards, the child groups cards together and then finds the appropriate label card for the set. Choose a card, such as the 2 of hearts, and ask the child to find a card that goes with it. There are many possibilities for the child. She may pick another 2, another heart, or another red card. An interaction between you and the child might go something like this:

ADULT: "Can you find a card that goes with this one?" (2 of hearts)

CHILD: (Places the 3 of hearts on the tray next to the chosen card.)

ADULT: "Can you find *another* card to go with these cards?" (There are still several possibilities for the child: any other heart card or any other red card will do.)

CHILD: (Places the 3 of diamonds with the other two cards.)

ADULT: "Why do these cards go together?" or "How are these cards alike?"

CHILD: "They are all red."

ADULT: "Can you find any other cards that go with these?"

The child will easily find other red cards to go with the cards already on the tray. This would be a good time to show her the "red" label card. Encourage the child to find "all the red" cards and place them with the others. This activity can be repeated many times by starting with different playing cards and making new sets.

Making Sets

DIRECTIONS: Turn the 10 label cards face down on the table. Place the 16 playing cards in a paper bag. The child is to pick one of the label cards, turn it face up, and place it on the tray. Child and adult then take turns picking a playing card from the bag and placing it either inside the set, on the tray, or outside the set, off the tray. All the cards can be placed either inside or outside the set. When they have all been placed, ask the child to name the set. This procedure can be repeated many times, starting with different label cards.

In a classroom, distribute all 16 cards, one to a child. Students can decide whether their cards go inside or outside the classroom set.

DIRECTIONS: Put the set of 16 playing cards into a paper bag. The child reaches in without looking and draws out 2 cards. She is to examine the 2 cards and answer the following questions:

How are these 2 cards *alike*?

How are these 2 cards *different*?

For example, the 2 cards could be the 2 of clubs and the 5 of spades. They are *alike* because they are the same color; they are *different* because they are different numerals and different suits. The 3 attributes for determining whether the cards are alike or different are color, number, and suit. In a classroom, this activity could be done individually by each child. Sets of cards could be made out of paper so that they can be pasted onto a recording sheet like the following:

FIGURE 7–12.

DIRECTIONS: This is a two-person game. The first player chooses a label card but doesn't let the other player see it. The second player tries to identify the label card in as few turns as possible.

Player one holds the label card in front of her so that she can refer to it. Player two picks up any playing card and asks, "Does this go inside or outside the set?" Player one places the card in the appropriate place, either on the tray or off the tray. The first turn, of necessity, is a wild guess; after that, logical reasoning skills are utilized. A card that goes outside the set gives many clues. An effort should be made to emphasize that cards placed outside give as much help as cards placed inside the set. It is not wrong or negative to choose a card that goes outside the set. In fact, often it is a deliberate choice.

Following is a sample game with some explanation of the clues given by each guess. Young children can often play this type of game but cannot verbalize what they are doing or why they are choosing a particular card. Even though they may not be able to verbalize it, they are, in fact, using logical reasoning.

Sample Game of *Guess My Set*

1st card: 2 of spades; goes inside the set.

Reasoning: the card is inside because it is a 2, a spade, or black.

2nd card: 2 of hearts; goes outside the set.

Reasoning: the label card cannot be the numeral 2, since the 2 is outside the set.

3rd card: 5 of spades; goes inside the set.

Reasoning: the label card is either the spade or black.

4th card: 5 of clubs.

Reasoning: if the card goes outside the set, the label card must be the spade; if the card goes inside the set, it must be black.

The 5 of clubs is placed outside the set. The label card must be the spade.

The child might now know that the label card is spades. If not, she can continue choosing cards, getting more clues, until she knows the answer and is sure of it.

Players should alternate being player one and player two.

FIGURE 7-13.

Guess My Card

SKILL AREA: Logical Reasoning (Deductive Reasoning)

GRADE LEVEL: Primary, intermediate

STRATEGIES: Guessing and testing
    Using manipulatives
    Success for each child
    Many approaches
    Game playing

MATH CONCEPTS: This activity provides practice in counting and recognition of numerals and shapes.
   The number concepts of *odd* numbers, *even* numbers, *greater than, less than, prime* numbers, *multiples,* and *divisible by* can be utilized in this game.

153

MATERIALS: A complete deck of playing cards (omitting the picture cards for younger children)

DIRECTIONS: The leader in the game chooses a card, records it, and returns it to the deck. Players are to discover the card by means of asking questions that can be answered "yes" or "no." After the leader answers a question, the player removes the cards which can no longer be considered. For example, if the question is "Is it red?" and the answer is "no," then all the red cards may be removed from the table. The leader should record the total number of questions asked. The object of the game is to guess the card by asking as few questions as possible. The questions and math concepts used depend on the age and ability of the children. At home this can be played as a two-person game. Some appropriate questions:

Is it red? Is it black?

Is it a heart? Diamond? Club? Spade?

Is it greater than _____? Is it less than _____?

The words *greater than* and *less than* do not include the number itself; for example, if the question is "Is it greater than 5?" is answered by "yes," then all the 5s, 4s, 3s, 2s, and aces could be removed from the deck.

Is it odd? Is it even?

Is it a multiple of 2? of 3? of 4? or 5?

Is it divisible by 2? by 3? by 4? by 5?

Is it *prime*?

The *prime* numbers in a deck of cards are 2, 3, 5, and 7. These are the numbers that are divisible only by themselves and 1.

Is it a picture card?

If picture cards are used, children will need to discriminate closely among them. For example:

Is it male?

Is it female?

Is there one eye showing?

Logical Reasoning

Hex and Nim are popular strategy games with simple rules that children can easily learn to follow. After playing the games several times, the players begin to develop strategies for winning.

SKILL AREA: Logical Reasoning (Deductive Reasoning)

GRADE LEVEL: Primary, intermediate, junior high

STRATEGIES: Game playing
  Using manipulatives
  Guessing and testing
  Many approaches

MATH CONCEPT: Simplifying the problem is a problem-solving strategy.

Hex[1]
(invented in the 1940s by Danish engineer Piet Hein)

MATERIALS: A hexagon-patterned playing board, two different colored crayons

FIGURE 7-14.

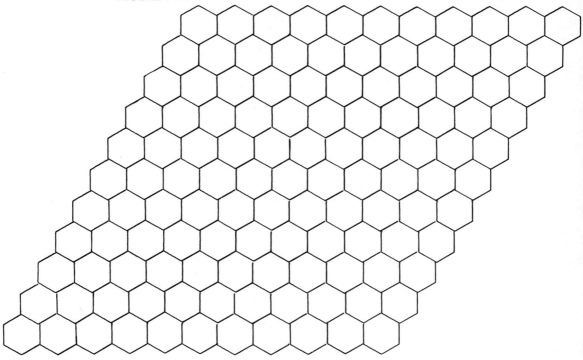

---

[1]Martin Gardner, *The Scientific American Book of Mathematical Puzzles and Diversions* (New York, N.Y.: Simon and Schuster), pp. 73–74. Copyright © 1959 by Martin Gardner.

DIRECTIONS: This is a game for two players. One player tries to make a connected path from the top to the bottom of the playing board. The other player tries to make the connection from side to side. The players take turns marking an X on any hexagon. The first player to make a completed path is the winner.

Nim

MATERIALS: 12 small objects, such as buttons, toothpicks, or beans

DIRECTIONS: Two players take turns playing this game. On each turn the player may remove 1, 2, or 3 objects from the group. The person who removes the last object is the winner. The number of objects may be increased when the players are no longer challenged by 12.

Both Hex and Nim can be simplified. Hex can be played with a smaller gameboard, such as 4 by 4.

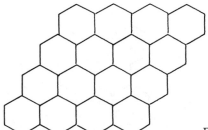

FIGURE 7-15.

Nim can be played with fewer objects. Simplifying these games facilitates the discovery of strategies to win. Simplification is helpful in many different kinds of math problems. With young children, you might want to start with simplified forms of the games and work up to the standard versions.

A Shopping Spree

SKILL AREA: Logical Reasoning (Combinatorial Reasoning)

GRADE LEVEL: Intermediate, junior high

STRATEGIES: Many right answers
Many approaches
Success for each child
Cooperative work
Using manipulatives
Estimating

MATH CONCEPTS: Computation, estimating answers
         Looking for all possibilities is a problem-solving strategy.

MATERIALS: Copies of the problems for each student; pencils;
paper; calculators, if available; cards with the following words
on them: SCARF, BELT, SWEATER, DRESS, SHOES, EARRINGS

DIRECTIONS: Have students work in groups to solve these problems. Distribute the problems in part one and a set of cards to
each student. Give students time to read the problems and ask
any questions to clarify the meaning. Then have them work on
problem one for a short time (5 to 10 minutes) to name as many
pairs of items as they can. Moving the cards around will help
them find all the possible pairs of items. One person in each
group should record the results and report them to the whole
class. After the results from all groups have been listed, ask if
there are any pairs of items that have not been found. Urge
students to work in their groups to try to organize the information so that they know for sure they have found all the possibilities. Have the groups share their methods for organizing the
information. Use the same procedure for problem two, where
they must find all the possible combinations of 3 different
items.

The problems in part two may be worked on in groups
also. Students may make and use cards if they wish, or they may
try these without using cards. Since these problems include
prices, students need to compute to find particular pairs or
groups of items. Allow discussion of the various ways they
organized this information and computed answers. Many of the
computations can be estimated to determine whether they are
larger or smaller than the answer sought. Exact answers are not
always necessary in these problems.

Problems

Part I

   SCARF   BELT   SWEATER   DRESS   SHOES   EARRINGS

1. Suzanne is going shopping for clothes. She can buy any 2
   different items. What can she buy?

2. Suppose that she can buy any 3 different items. What can
   she buy?

Part II

1. Debbie is going shopping for clothes. She has saved exactly
   $32 and she wants to spend all of her money. If she buys 2

different items, what can she buy? If she buys 3 different items, what can she buy?

SLIPPERS	HAT	BLOUSE	BOOTS	SUIT	BRACELET
$5	$13	$14	$19	$27	$8

2. Debbie brought her friend Alice back the next week to the same store and found that the prices had changed. Alice has saved $45 and wants to spend it all. If she buys 2 different items, what can she buy? What 3 different items can she buy?

SLIPPERS	HAT	BLOUSE	BOOTS	SUIT	BRACELET
$8	$9	$17	$36	$19	$28

Four Scientists

SKILL AREA: Logical Reasoning (Deductive Reasoning, Combinatorial Reasoning)

GRADE LEVEL: Intermediate, junior high

STRATEGIES: Many approaches
       Cooperative work
       Content relevance
       Guessing and testing
       Modeling new options

MATH CONCEPTS: Looking for all the possibilities is a problem-solving strategy.
    Analyzing clues is a problem-solving strategy.
    Making a diagram is a problem-solving strategy.
    Finding contradictions in information is an important part of mathematical thinking.

MATERIALS: A copy of the problem for each student, pencils, paper

DIRECTIONS: Have students work on this problem in groups. Distribute the problem, and allow time for the students to read it and ask questions.

Problem

Four women—Alice, Barbara, Cheryl, and Diane—are scientists. Their occupations are achaeologist, botanist, chemist, and doctor, but not necessarily in that order. One day they were seated around a square table. From the following clues, can you determine which person has which occupation?

1. The person who sat across from Barbara was the chemist.

2. The person who sat across from Diane was not the archaeologist.

3. The person who sat on Alice's left was the doctor.

4. The person who sat on Cheryl's left was not the botanist.

5. Only one person's name starts with the same letter as her occupation.

6. The archaeologist and botanist are sitting next to each other.

ANSWER: Alice, chemist; Barbara, archaeologist; Cheryl, botanist; Diane, doctor

If students need help getting started, you might draw large squares on the chalkboard to represent the square table in the problem. Draw as many squares as there are groups. Have a maximum of 6 groups, since there are 6 different possible ways that the women can be seated. Ask for suggestions of ways that the scientists might be seated around the table, and write their names in the appropriate places on the squares. When all the squares have names on them, assign one square to each group. Urge students to read all the clues carefully to determine if they have an arrangement that works, that is, that there are no contradictions. Have each group explain to the others why their arrangement works or doesn't work. If none of the arrangements works, ask students to find other possible ways that the women might be seated. Suggest that they organize their information so that they know when they have found all the ways. Then have them test the remaining possibilities to find a solution.

8 Scientific Investigation

What is science? Science is a way of predicting events. Understanding scientific investigation is central to understanding science. Through the investigative process, scientists generate hypotheses and determine their predictive power. A hypothesis becomes part of our established body of knowledge after many experiments have demonstrated that it leads to valid predictions.

In conceiving and implementing a research project, a scientist uses many different investigative skills. These include:

1. making observations,

2. asking questions,

3. formulating and assessing hypotheses,

4. designing experiments,

5. collecting and analyzing data,

6. drawing conclusions, and

7. communicating results.

The importance of introducing investigation early and incorporating it into the curriculum at every level cannot be overemphasized, because in addition to their applications to the various fields of science, investigative skills are useful in everyday life.

All the reasoning skills discussed in chapter seven are used in scientific investigation. For example, control of variables is used in designing experiments, probabilistic reasoning is used in selecting and applying statistical tests to analyze data, and deductive reasoning is used in drawing conclusions. Investigative skills combine diverse thinking skills, experience, and knowledge.

Even very young children can carry out simple investigations. When investigative skills are developed gradually, through many experiences at home and school, they become easy and natural. Scientific investigation will never seem difficult or mysterious to a girl with strong component skills built through concrete experiences.

The first two activities in this chapter, Asking About Animals and Patterns in Nature, emphasize the first three steps of the investigative process: making observations, asking questions, and formulating hypotheses. In these activities, children observe animals or biological communities, ask questions based on their observations, and guess answers to their questions.

Robert Knott[1] and others have found that girls tend to formulate fewer hypotheses, that is, make fewer guesses, than boys. Girls' inhibition in formulating hypotheses is probably due to the risk involved: the questions might be foolish; the hypotheses might be wrong. Asking About Animals and Patterns in Nature reinforce children for taking risks while helping them to develop important investigative skills.

Heart Rate, Making Mountains, Plant Growth, and Mealworm Behavior emphasize designing experiments, collecting and analyzing data, and drawing conclusions. Perhaps the most important skills that these activities help to develop are identification and control of variables. Boys' toys and games, which often involve construction and experimentation, are conducive to the development of these skills. Our activities give girls as well as boys the opportunity to perform controlled experiments.

Paper Airplanes requires children to design and test paper airplanes to see which one will fly the farthest. In doing so, they gain experience in hypothesis formulation and the subsequent

---

[1]Robert Church Knott, "Hypothesis Generation, Selection of Hypotheses, and Experimental Designs by Individuals and Groups of Children in Grades Four, Six, and Eight" (unpublished doctoral dissertation, University of California at Berkeley, 1972), pp. 88–91.

steps of the investigative process. Some children may even bring the entire investigative process into play, because their observations can lead to additional questions, such as "What makes some planes fly higher than others?" and "What makes some planes turn?" Since folding a paper airplane is a spatial task, this activity also helps to develop spatial visualization skills. The activity can be a thrilling experience for girls, because it gives them an opportunity to be successful on a task (paper airplane making) on which boys are usually more successful due to past experience.

Asking About Animals

SKILL AREA: Scientific Investigation (Making Observations, Asking Questions, and Formulating Hypotheses)

GRADE LEVEL: Primary, intermediate, junior high

STRATEGIES: Success for each child
Many right answers
Independent work
Content relevance

SCIENCE CONCEPTS: An *adaptation* is a characteristic that helps a plant or animal to survive in its environment.

Adaptations help animals carry out their basic life functions (feeding, locomotion, reproduction, maintenance of body temperature and fluids, avoidance of predators).

MATERIALS: Paper, pencils, crayons, pictures of animals

DIRECTIONS: If possible, this activity should coincide with a visit to a zoo. It can also be done indoors, using pictures of animals. As the children look at animals, give them examples of questions that refer to animal adaptations. Some sample questions:

Why does the duck have webbed feet?

Why does the polar bear have white fur?

Why does the zebra have stripes?

Why does the vulture have a hooked beak?

Why do male red-winged blackbirds have red spots on their wings?

Why do mussels have shells?

Give examples of answers that emphasize survival in the environment: the duck has webbed feet to help it swim; the vulture has a hooked beak to help it tear meat.

Ask the children for additional questions. Have them guess answers to some of the questions. (Guesses that attempt to answer the sample questions are given at the end of this activity.) Emphasize that questions and guesses are just as important as answers and that a question might have more than one right answer.

If you are working with primary children, have each child (1) think of a question based on her own observations of animals, (2) draw a picture to illustrate the question, (3) guess answers to the question, and (4) write a story that includes an answer to the question. Very young children can tell their stories orally.

Have older children pick one characteristic of animals, such as color, kind of forelimb, or shape of mouth or beak, and compare four or more animals that differ with respect to this characteristic, speculating about the reasons for the differences. Encourage children to rely on their own thinking and make guesses, rather than limiting themselves to "right" answers.

Some sample questions and guesses:

Why does the duck have webbed feet?
—To help it swim
—To help it walk on sand

Why does the polar bear have white fur?
—So that its enemies will have trouble seeing it in snow
—So that its prey won't see it approaching in snow

Why does the zebra have stripes?
—So that enemies crouched in the grass will have trouble seeing it
—To help zebras recognize each other

Why does the vulture have a hooked beak?
—To tear meat
—To attack enemies

Why do male red-winged blackbirds have red spots on their wings?
—To attract mates
—To intimidate enemies

Why do mussels have shells?
—To protect them from enemies
—To help them stay moist at low tide
—To help them stay cool at low tide

SKILL AREA: Scientific Investigation (Making Observations, Asking Questions, and Formulating Hypotheses)

GRADE LEVEL: Primary, intermediate, junior high

STRATEGIES: Success for each child
　　　　　　Many right answers
　　　　　　Independent work
　　　　　　Cooperative work
　　　　　　Content relevance

SCIENCE CONCEPTS: Plants and animals are *distributed* unevenly in nature. For example, there might be many bay trees at two locations in a park and none elsewhere in the park.

Many factors affect plant and animal distribution. These factors can be *physical,* such as light, wind, temperature, moisture, and soil conditions. They can also be *biological*; for example, predators or prey.

MATERIALS: Paper, pencils, crayons

DIRECTIONS: This activity should begin during a visit to a park. Call children's attention to patterns in plant and animal distribution, and give them examples of questions that ask why these patterns exist. Some sample questions:

Why do ferns grow on only one side of the creek?

Why don't any plants grow under these eucalyptus trees?

Why are there more dragonflies in sunny areas than in shady areas?

Why are lichens growing on the cypress trees but not on the oaks?

Why are those little gray birds on the ground instead of in the trees?

Ask children to guess possible answers to your questions and to look for additional patterns.

Back inside, have children brainstorm answers to some of your sample questions. If necessary, prompt them to increase the number of guesses. Explain that many factors affect plant and animal distribution and that more than one of the guesses for each question might be correct. Distinguish between physical and biological factors.

Primary children should (1) draw a picture showing a pattern observed outdoors (not using your examples, if possible), and (2) write down a question based on the pattern and a possible answer. Have the children share their pictures, questions, and guesses.

Have older children (1) write down a question based on a pattern observed outdoors (not using your examples, if possible), and (2) list as many guesses as possible that might answer the question. Let children discuss their questions and guesses in groups of 2 to 4. As a result of the discussion, each child should put a star by each guess she thinks is especially likely to be correct. Have children share their questions and starred guesses with the class, explaining why they think these guesses are good ones. Emphasize that there are *no wrong answers* in this activity: the purpose is to practice thinking like scientists.

Some sample guesses:

Why do ferns grow on only one side of the creek?
—The other side is too bright for them.
—The other side is too dry for them.
—The other side is too hot for them.
—Ivy crowds them out on the other side.
—The soil on the other side contains a chemical that kills them.
—The soil on the other side lacks a nutrient that they need.

Why don't any plants grow under these eucalyptus trees?
—The trees block out too much light.
—The trees produce a poison that kills other plants.
—The trees use up all the nutrients in the soil.
—The trees use up all the water in the soil.

Why are there more dragonflies in sunny areas than in shady areas?
—They're attracted to light.
—They're attracted to heat.
—They're attracted to dryness.
—Their food is found in sunny areas.

Why are lichens growing on the cypress trees but not on the oaks?
—Oak bark contains a chemical that kills them.
—Cypress bark contains a nutrient they need.
—Cypress bark provides a better surface for them to attach to.
—There's more moisture in the cypress bark.

Why are those little gray birds on the ground instead of in the trees?

—They eat insects found on the ground.

—They eat seeds found on the ground.

—Other kinds of birds are taking up the space in the trees.

—They're collecting material to build nests.

—They live near the ground, not up in the trees.

Heart Rate

SKILL AREA: Scientific Investigation (Making Observations, Controlling Variables, and Drawing Conclusions)

GRADE LEVEL: Primary, intermediate

STRATEGIES: Success for each child
Many right answers
Using manipulatives
Cooperative work
Content relevance

SCIENCE CONCEPTS: *Heart rate* is the number of times our heart beats each minute.

Each time our heart beats, it sends blood through vessels called *arteries*. This makes the arteries stretch a little bit. We can feel the stretching of an artery as our *pulse*.

We can measure our heart rate indirectly by counting the pulse at our wrist.

Our heart rate increases when we exercise.

MATERIALS: Graph paper; pencils; stopwatch, timer clock, or watch with second hand

DIRECTIONS: Working in pairs, have the children practice finding the pulse on each other's wrist. When they can do this, begin the activity.

Children will measure each other's pulse after engaging in the following activities: lying down, sitting, walking, running, jumping. They should do the activities in the order given, engaging in each one for 3 minutes before taking their pulse. They should count the pulse for 15 seconds. Signal them to let them know when to start and stop counting. To avoid confusion, have the two children in each team measure each other's pulse separately, rather than simultaneously. Ask them to write down their measurements and save them to make a graph.

Have each child make a bar graph to illustrate her heart rate. (See chapter five for activities to introduce graphing.)

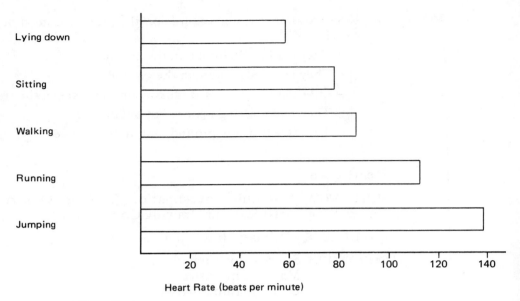

Lying down

Sitting

Walking

Running

Jumping

Heart Rate (beats per minute)

FIGURE 8-1.

Older children should multiply the 15-second rate by 4 and plot beats per minute. They can also compute averages based on class data for each activity.

Making Mountains

SKILL AREA: Scientific Investigation (Making Observations, Formulating Hypotheses, Controlling Variables, and Drawing Conclusions)

GRADE LEVEL: Primary, intermediate

STRATEGIES: Success for each child
Using manipulatives
Cooperative work

SCIENCE CONCEPTS: Rocks form more stable structures than either soil or sand.

Neither dry soil nor dry sand can be built up very high, because it will slide or blow away.

Damp soil and damp sand can be built up high, but the addition of more water will make them slide.

In nature, when soil or sand slides or blows away, the process is called *erosion*.

MATERIALS: Paper cups, soil, sand, small rocks, water, rulers, graph paper, pencils, paper plates or disposable meat trays,

168

newspapers (to protect the area where children work), large containers for holding dry materials and disposing of wet materials

DIRECTIONS: Show children how to make a mountain by filling a paper cup with soil, sand, or rocks. Tell them that their task will be to make different kinds of mountains, using the material available, and to measure the peaks:

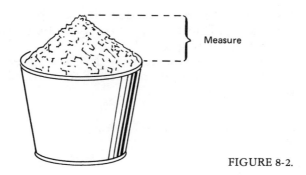

Measure

FIGURE 8-2.

They should make as many different kinds of mountains as they can. It's okay to mix water with the soil, sand, or rocks, but not to mix these materials with each other.

Have children work in pairs. Paper plates or meat trays placed beneath the cups will facilitate clean-up later. Most teams will make and measure the following kinds of mountains: rock, dry sand, damp sand, dry soil, damp soil. Tell children to write down their measurements because they will use them later to make a graph.

After children have had time to work on their own, ask them to try the following experiment:

1. Make a rock mountain. Pour water on the mountain.

2. Make a mountain with damp soil or sand. Pour water on the mountain.

One or two rocks may tumble from the rock mountain, but the structure will stay mostly intact. The soil or sand mountain can be completely washed away. (The same amount of water should be poured in steps 1 and 2.)

Pose the following questions. Let children provide the answers.

1. Can a very high mountain be made only of soil or sand? Why or why not? (No. Wind, water, and gravity would quickly wear it away.)

2. Plant roots hold soil in place by optimizing water content. What would happen if all the plants on a hill were destroyed, as by fire or land clearing? (In the dry season, the soil would slide or blow away; in the wet season it would slide.)

Use the term *erosion* to describe the processes by which land is worn away. Point out that rock can be eroded, too, but that the process is more gradual.

Conclude the activity by having each pair make a bar graph to illustrate its results. (See introductory graphing activities in chapter five.)

FIGURE 8-3. Graph showing heights of various kinds of mountains made in a paper cup.

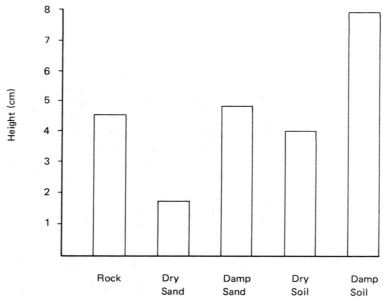

Seed Germination and Plant Growth

In these activities, children will plant seeds in paper cups to determine the requirements of seed germination and plant growth. At home, your child can do a series of experiments to test each of the relevant factors.

SKILL AREA: Scientific Investigation (Formulating Hypotheses, Making Observations, Designing Experiments, Controlling Variables, and Drawing Conclusions)

GRADE LEVEL: Primary, intermediate

STRATEGIES: Success for each child
Many right answers
Many approaches
Guessing and testing
Using manipulatives
Cooperative work
Content relevance

SCIENCE CONCEPTS: In order to sprout, seeds need moisture, air, and warmth. In addition, they should not be planted too deep in the soil.

Plant growth depends on moisture, air, temperature, soil, nutrients, and light.

Seed Germination

MATERIALS: Beans, soil, paper cups, paper plates, water, plant food, paper towels, jar lids, plastic bags, masking tape

DIRECTIONS: Ask children what they think seeds need in order to sprout. They should generate a list that includes the following possibilities: moisture, air, warmth, soil, nutrients, light, proper depth in soil.

Have the children work in pairs. Each pair will perform an experiment to test one of the factors on the list. Every experiment should have at least 2 conditions in order to make a fair test. For example, children testing moisture should plant seeds in 2 cups: one cup should be watered, while the other is kept dry.

Younger children will need a lot of guidance. Have a discussion in which the children suggest experiments. Ask questions and give hints to help them refine their ideas. Let older children design their own experiments, helping them as necessary.

Planting instructions: Have children plant 2 seeds in each cup. The seeds should be approximately ½ in. (1.25 cm) deep in the soil, except when seed depth is being tested. Each cup should have a small hole in the bottom, so that excess water can run out. Put paper plates under the cups. All cups should be labeled with children's names and experimental conditions.

Possible experiments:

Moisture. Plant seeds in each of 2 cups. Water only one cup.

Air. Place 2 seeds in a cup of water. These seeds have no air, but also no soil. To make a fair test, comparison seeds should have air, but no soil: Place seeds on a piece of moistened paper towel in a jar lid. Put the lid into a plastic bag to slow evaporation. Make pinholes in the bag to allow gas exchange.

Warmth. Plant seeds in each of 2 cups. Put one cup in the refrigerator and one in a cabinet. The comparison should be in a cabinet so that neither cup will have any light.

Soil. Try sprouting seeds in a jar lid, as described above for the air experiment. As a comparison, plant seeds in soil.

Nutrients. Plant seeds in each of 2 cups. Give plant food only to one.

Light. Plant seeds in each of 2 cups. Put one cup near a window and one in a cabinet or under a box.

Depth in soil. In different cups, plant seeds at several different depths.

The soil should be kept moist in all cups except the moisture comparison, and the paper towels should be kept moist in the jar lids. Put all cups and lids near a sunny window, unless otherwise indicated. Allow two weeks to determine which factors are necessary for germination.

Have a discussion of the results. Be sure that all children have the opportunity to observe the results of all the different experiments.

Plant Growth

MATERIALS: Beans, soil, paper cups, paper plates, water, plant food, paper towels, jars, plastic bags, string, masking tape, graph paper, rulers

DIRECTIONS: Ask children what they think plants need in order to grow. They should generate a list that includes the following factors: moisture, air, warmth, soil, nutrients, and light.

Have children work in pairs. Each pair should test one of the factors on the list. To begin the activity, they should plant 2 seeds in each of 2 cups, following the planting instructions given under Seed Germination. (You might want to plant extra seeds in case some of them do not come up.) The cups should be kept near a sunny window and watered regularly.

When the plants are about 1½ in. (4cm) high, begin the experiments, which should be similar to the seed germination experiments. Every experiment should have both an experimental condition and a comparison. If necessary, nutrients can be given to all the plants except the comparisons in the nutrient experiments. Soil can be tested by removing one of the plants from its cup, carefully rinsing the roots, and putting it in a jar with a moistened paper towel crumpled at the bottom. Air can be tested by putting a small plastic bag over a plant and securing it with a string just above the soil.

Younger children will need a lot of help in designing their experiments. Older children should need less help. Give hints only if they have trouble coming up with a reasonable plan.

The plants should be measured the day that the experiments begin and at least once a week for the following 3 weeks. If both seeds in a cup sprout, one should be selected to use in the experiment. At the end of 3 weeks, have each child make a line graph showing the growth of the experimental and comparison plant (see introductory graphing activities in chapter five):

Have a discussion of the results. Be sure that all children

FIGURE 8-4. Graph showing the growth of two plants, one kept in darkness and the other receiving light.

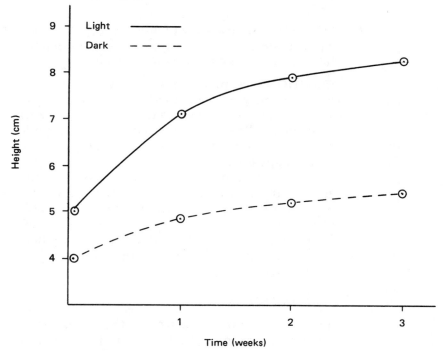

have the opportunity to observe all the different experiments and to look at all the different graphs.

Mealworm Behavior

SKILL AREA: Scientific Investigation (Formulating Hypotheses, Making Observations, Designing Experiments, Controlling Variables, and Drawing Conclusions)

GRADE LEVEL: Primary, intermediate

STRATEGIES: Success for each child
　　　　　　Many right answers
　　　　　　Many approaches
　　　　　　Guessing and testing
　　　　　　Using manipulatives
　　　　　　Cooperative work

SCIENCE CONCEPTS: A mealworm is the larva (plural *larvae*) of a mealworm beetle.
　　　A mealworm eats cereal.
　　　A mealworm responds to environmental factors, such as moisture and light.
　　　If a mealworm has food, moisture, air, and warmth, it will change first into a pupa (plural *pupae*) and then into an adult beetle.

MATERIALS: Mealworms (available at pet stores, about 2¢ each), paper towels, water, oatmeal, shredded paper, plastic wrap, shoe boxes, masking tape, cheesecloth

DIRECTIONS: Explain that mealworms are not really worms: they are the young of mealworm beetles. Introduce the word "larva." Ask children what they think mealworms need in order to survive. Their suggestions should include air, food, and moisture. Appropriate light and temperature might be additional suggestions.
　　　Tell children that they will do experiments to determine if, during a period of 24 hours, mealworms will move toward light, food, or moisture. Have them work in groups of 4. Each group should test one factor. At home, your child can test all 3 factors. Younger children will need a lot of guidance. Older children should design their own experiments. Help them only as necessary.

Possible experiments:

Light. Put 5 or 6 mealworms in the center of a shoe box. At one end of the box put a crumpled paper towel or some shredded paper. Cover the box with plastic wrap. Poke holes in the plastic wrap. Do the mealworms prefer light or shade?

Food. Put 5 or 6 mealworms in the center of a shoe box. At one end of the box put some oatmeal, at the other end some shredded paper. Cover the box with plastic wrap. Poke holes in the plastic wrap. Do the mealworms prefer the oatmeal to the paper?

Moisture. Put 5 or 6 mealworms in the center of a shoe box. Crumple a moistened paper towel at one end of the box, a dry paper towel at the other. Cover the box with plastic wrap. Poke holes in the plastic wrap. Which end do the mealworms prefer?

All the boxes should be labeled with masking tape to identify the investigators and the type of experiment. Have children start the experiments in the morning and check them that afternoon and again the following morning. Be sure that all children have the opportunity to observe all 3 types of experiment. Have a discussion of the results.

In order to *metamorphose* (change into beetles), mealworms need air, food, moisture, and the right temperature. You can easily make them metamorphose by doing the following:

> Fill a bowl two thirds with oatmeal. On top of the oatmeal place an apple or potato slice; a piece of crumpled, moistened paper towel; and about 10 mealworms. Cover the bowl with cheese-cloth, and tape the cheesecloth to the sides of the bowl. Place the bowl in a warm, but not hot, area of the room.

Check the mealworms twice a week. Use an eyedropper to keep the paper towel moist. From time to time, add a fresh apple or potato slice, a fresh paper towel, and fresh oatmeal. The mealworms will change first into pupae, then into beetles. (Pupae are shorter, fatter, and lighter in color than larvae, and they do not crawl.) The entire process should take 2 to 6 weeks.

Have children observe and describe the animals at each stage of development (larvae, pupae, beetles). What color are they? Do they have eyes? How long are they? How do they move? Many different kinds of insects, including flies, wasps, and butterflies, have the same kind of life cycle as the mealworm beetle.

Paper Airplanes

SKILL AREA: Scientific Investigation (Formulating Hypotheses, Making Observations, Designing Experiments, Controlling Variables, and Drawing Conclusions)

GRADE LEVEL: Intermediate, junior high

STRATEGIES: Using manipulatives
Success for each child
Many right answers
Many approaches
Guessing and testing
Cooperative work
Single-sex and mixed-sex groups

SCIENCE CONCEPTS: A paper airplane receives *energy* from the arm of the person who throws it.

Friction is produced when two substances rub against each other. A paper airplane loses energy in the form of friction by rubbing against molecules of air.

The plane producing the least friction will fly fastest.

The plane with the most resistance to vertical motion will stay in the air the longest.

The plane that flies the farthest is not necessarily either the one that flies the fastest or the one that stays in the air the longest.

MATERIALS: Several sheets of paper (8½ in. by 11 in. or 20cm by 30cm) for each child, tape, paper clips

DIRECTIONS: Have children work in same-sex pairs at the beginning of this activity. Each pair should make 3 to 5 different kinds of paper airplanes. Show children how to fold 2 different kinds to get them started.

The Drifter

1. Fold the paper in half lengthwise.

2. Fold over two corners (separately, in opposite directions) to meet the center line.

3. To make wings, fold each side to meet the center line.

4. Open the wings, secure with tape, and it's ready to fly!

FIGURE 8-5. The Drifter.

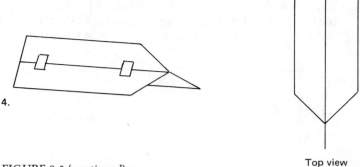

4.

FIGURE 8-5 (continued).

Top view

The Arrow

Steps 1 and 2 are the same as for the Drifter.

3. Starting at the nose, fold each side partway to the center line.

4. Starting at the nose, fold each side to meet the center line.

5. Open the wings, secure with tape, and it's ready to fly!

FIGURE 8-6. The Arrow.

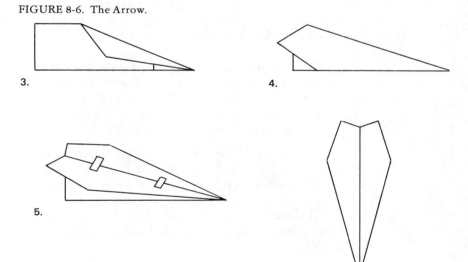

3.

4.

5.

Top view

Possible variations of these planes: For a narrower plane, make additional folds in the wings. To weight the plane at the front, make several ¼-in. folds at one end of the paper *before* folding the plane.

There are many other ways to fold a paper airplane. Planes can differ in length, wingspan, weight distribution, and shape of wings. Here are some additional wing shapes:

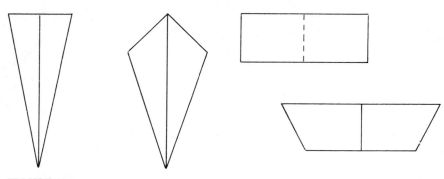

FIGURE 8-7.

Outdoors or in an auditorium or gym, have each pair of children test their planes to see which one will fly the farthest. All children should observe the testing. At this stage, do not have the teams compete against each other.

After the testing, hold a "scientific conference" to establish the most important features of a good paper airplane. Children can vote on whether various features are beneficial or detrimental to flight. During the discussion, the following points should be made:

> *Stable* planes tend to fly farther than unstable ones. Stability is increased by *neat, symmetrical folding* and by *weighting the plane at the front*. The effects of weighting can be demonstrated by adding two paper clips to the front of a plane. Try putting the clips instead on the middle or back. What happens?

> *Streamlined* planes fly farther than unstreamlined ones. A streamlined plane has long, smooth lines past which air can flow with little resistance. The Arrow is more streamlined than the Drifter.

If time permits, children can also vote on the characteristics of the plane that flies the highest. In this case, *wing flaps* should be mentioned:

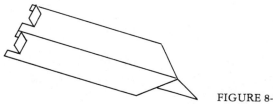

FIGURE 8-8.

If flaps are pointed upward, the plane will tend to go up, because air will push the tail down. The opposite occurs if the flaps are pointed down. If one flap is up and the other is down, the plane will spin.

After the discussion, hold a contest to see who can make the plane that will fly the farthest. Have children work in pairs (mixed-sex pairs okay). Each pair should enter 2 planes and keep the record of the one that flies the farthest.

Resource
List

The following books and materials contain a variety of hands-on activities that can help girls develop critical math and science skills.

MATHEMATICS

Askew, Judy. *The Sky's the Limit in Math-Related Careers*. U.S. Department of Education: Women's Educational Equity Act Program, 1981.

Baratta-Lorton, Mary. *Workjobs . . . For Parents*. Menlo Park, Calif.: Addison-Wesley Publishing Co., 1972.

Barnett, Carne. *Metric Ease*. Palo Alto, Calif.: Creative Publications, 1975.

Bezuzska, Stanley; Kenney, Margaret; and Silvey, Linda. *Tessellations: The Geometry of Patterns*. Palo Alto, Calif.: Creative Publications, 1977.

Brydegaard, Marguerite, and Inskeep, James E. Jr., Eds. *Readings in Geometry from the Arithmetic Teacher*. Washington, D.C.: National Council of Teachers of Mathematics, 1970.

Burns, Marilyn. *The Book of Think*. Boston: Little Brown and Company, 1976.

Burns, Marilyn. *The I Hate Mathematics Book*. Boston: Little, Brown and Company, 1975.

Downie, Diane; Slesnick, Twila; and Stenmark, Jean K. *Math for Girls and Other Problem Solvers*. Berkeley, Calif.: Regents, University of California, 1981.

Kaseberg, Alice; Kreinberg, Nancy; and Downie, Diane. *Use EQUALS to Promote the Participation of Women in Mathematics*. Berkeley, Calif.: Math/Science Network, Lawrence Hall of Science, University of California, 1980.

Meiring, Steven P. *Parents and the Teaching of Mathematics*. Columbus, Ohio: Ohio Department of Education, 1980.

Pedersen, Jean J., and Armbruster, Franz O. *A New Twist: Developing Arithmetic Skills Through Problem Solving*. Menlo Park, Calif.: Addison-Wesley Publishing Company, 1979.

Perl, Teri. *Math Equals: Biographies of Women Mathematicians + Related Activities*. Menlo Park, Calif.: Addison-Wesley Publishing Company, 1978.

Rand, Ken. *Point-Counterpoint: Graphing Ordered Pairs*. Palo Alto, Calif.: Creative Publications, 1979.

Seymour, Dale. *Tangramath*. Palo Alto, Calif.: Creative Publications, 1971.

Sharp, Evelyn. *Thinking is Child's Play*. New York: Avon Books, 1969.

Tobias, Sheila. *Overcoming Math Anxiety*. New York: W. W. Norton & Company, Inc., 1978.

Walter, Marion I. *Boxes, Squares, and Other Things: A Teacher's Guide for a Unit in Informal Geometry*. Reston, Va.: National Council of Teachers of Mathematics, Inc., 1970.

Zaslavsky, Claudia. *Preparing Young Children for Math: A Book of Games*. New York: Schocken Books, 1979.

SCIENCE

Books

Allison, Linda. *Blood and Guts*. Boston: Little, Brown and Co., 1976.

Allison, Linda. *The Sierra Club Summer Book*. San Francisco/New York: Sierra Club Books/Charles Scribner's Sons, 1977.

Amery, Heather, and Littler, Angela. *The FunCraft Book of Magnets and Batteries*. New York: Scholastic Book Service, 1976.

Beck, Derek, and McNeil, Mary Jean. *The Fun Craft Book of Flying Models: Paper Planes that Really Fly.* New York: Scholastic Book Service, 1976.

Cobb, Vicki. *More Science Experiments You Can Eat.* New York: J. B. Lippincott, 1979.

Cobb, Vicki. *Science Experiments You Can Eat.* New York: J. B. Lippincott, 1972.

De Caro, Matthew V. *The Gray's Anatomy Coloring Book.* Philadelphia: Running Press, 1980.

Goldstein-Jackson, Kevin; Rudnick, Norman; and Hyman, Ronald. *Experiments with Everyday Objects: Science Activities for Children, Parents, and Teachers.* Englewood Cliffs, N.J.: Prentice-Hall, Inc., 1978.

Herbert, Don. *Mr. Wizard's Supermarket Science.* New York: Random House, 1980.

McCoy, Elin. *The Incredible Year-Round Playbook: Fun with Sun, Sand, Water, Wind and Snow.* New York: Random House, 1979.

Nickelsburg, Janet. *Nature Activities for Early Childhood.* Reading, Mass.: Addison-Wesley, 1976.

Simons, Robin. *Recyclopedia: Games, Science Equipment, and Crafts from Recycled Materials.* Boston: Houghton Mifflin Co., 1976.

Van Deman, Barry A., and McDonald, Ed. *Nuts and Bolts: A Matter of Fact Guide to Science Fair Projects.* Harwood Heights, Ill.: The Science Man Press, 1980.

White, Lawrence B., Jr. *Science Toys and Tricks.* Reading, Mass.: Addison-Wesley, 1975.

Activity Packets

Buller, Dave; De Lucchi, Linda; Knott, Robert C.; and Malone, Larry. *Outdoor Biology Instructional Strategies* (OBIS). Nashua, N.H.: Delta Education, 1980.

Humbolt County Office of Education. *Green Box.* Eureka, Calif.: Humbolt County Office of Education, 1975.

Sly, Carolie, and Whitely, Molly. *Using Wild Edible Plants with Children.* Berkeley, Calif.: Instructional Laboratory, School of Education, University of California, 1979.

Steinberg, Barbara. *Spaceship School,* rev. ed., San Rafael, Calif.: Marin County Office of Education, 1982.

Index